Epigenetics to Optogenetics - A New Paradigm in the Study of Biology

Edited by Mumtaz Anwar, Zeenat Farooq, Riyaz Ahmad Rather, Mohammad Tauseef and Thomas Heinbockel

Published in London, United Kingdom

Epigenetics to Optogenetics - A New Paradigm in the Study of Biology
http://dx.doi.org/10.5772/intechopen.87476
Edited by Mumtaz Anwar, Zeenat Farooq, Riyaz Ahmad Rather, Mohammad Tauseef
and Thomas Heinbockel

Contributors
Fadime Eryılmaz Pehlivan, Mumtaz Anwar, Zeenat Farooq, Ambreen Shah, Mohammad Tauseef, Riyaz
Ahmad Rather, Mehmet Ünal, Sivasankari Sivaprakasam, Vinoth Mani, Nagalakshmi Balasubraniyan, David
Ravindran Abraham, Jun Tamogami, Takashi Kikukawa, Brent M. Michael Bijonowski, Thomas Heinbockel,
Rasha A. Alhazzaa, Antonei B. Csoka

Notice
Statements and opinions expressed in the chapters are these of the individual contributors and
not necessarily those of the editors or publisher. No responsibility is accepted for the accuracy
of information contained in the published chapters. The publisher assumes no responsibility for
any damage or injury to persons or property arising out of the use of any materials, instructions,
methods or ideas contained in the book.

First published in London, United Kingdom, 2022 by IntechOpen
IntechOpen is the global imprint of INTECHOPEN LIMITED, registered in England and Wales,
registration number: 11086078, 5 Princes Gate Court, London, SW7 2QJ, United Kingdom
Printed in Croatia

British Library Cataloguing-in-Publication Data
A catalogue record for this book is available from the British Library

Additional hard and PDF copies can be obtained from orders@intechopen.com

Epigenetics to Optogenetics - A New Paradigm in the Study of Biology
Edited by Mumtaz Anwar, Zeenat Farooq, Riyaz Ahmad Rather, Mohammad Tauseef
and Thomas Heinbockel
p. cm.

This title is part of the Biochemistry Book Series, Volume 35
Topic: Cell and Molecular Biology
Series Editor: Miroslav Blumenberg
Topic Editor: Rosa María Martínez-Espinosa

Print ISBN 978-1-83880-993-5
Online ISBN 978-1-83881-123-5
eBook (PDF) ISBN 978-1-83881-124-2
ISSN 2632-0983

IntechOpen Book Series
Biochemistry
Volume 35

Aims and Scope of the Series

Biochemistry, the study of chemical transformations occurring within living organisms, impacts all of the life sciences, from molecular crystallography and genetics, to ecology, medicine and population biology. Biochemistry studies macromolecules - proteins, nucleic acids, carbohydrates and lipids –their building blocks, structures, functions and interactions. Much of biochemistry is devoted to enzymes, proteins that catalyze chemical reactions, enzyme structures, mechanisms of action and their roles within cells. Biochemistry also studies small signaling molecules, coenzymes, inhibitors, vitamins and hormones, which play roles in the life process. Biochemical experimentation, besides coopting the methods of classical chemistry, e.g., chromatography, adopted new techniques, e.g., X-ray diffraction, electron microscopy, NMR, radioisotopes, and developed sophisticated microbial genetic tools, e.g., auxotroph mutants and their revertants, fermentation, etc. More recently, biochemistry embraced the 'big data' omics systems. Initial biochemical studies have been exclusively analytic: dissecting, purifying and examining individual components of a biological system; in exemplary words of Efraim Racker, (1913 –1991) "Don't waste clean thinking on dirty enzymes." Today, however, biochemistry is becoming more agglomerative and comprehensive, setting out to integrate and describe fully a particular biological system. The 'big data' metabolomics can define the complement of small molecules, e.g., in a soil or biofilm sample; proteomics can distinguish all the proteins comprising e.g., serum; metagenomics can identify all the genes in a complex environment e.g., the bovine rumen.

This Biochemistry Series will address both the current research on biomolecules, and the emerging trends with great promise.

Meet the Series Editor

Miroslav Blumenberg, Ph.D., was born in Subotica and received his BSc in Belgrade, Yugoslavia. He completed his Ph.D. at MIT in Organic Chemistry; he followed up his Ph.D. with two postdoctoral study periods at Stanford University. Since 1983, he has been a faculty member of the RO Perelman Department of Dermatology, NYU School of Medicine, where he is codirector of a training grant in cutaneous biology. Dr. Blumenberg's research is focused on the epidermis, expression of keratin genes, transcription profiling, keratinocyte differentiation, inflammatory diseases and cancers, and most recently the effects of the microbiome on the skin. He has published more than 100 peer-reviewed research articles and graduated numerous Ph.D. and postdoctoral students.

Meet the Volume Editors

Mumtaz Anwar, Ph.D., is currently a research assistant professor at the Department of Pharmacology and Regenerative Medicine, College of Medicine, the University of Illinois at Chicago, USA. Dr. Anwar obtained his Ph.D. in Cancer Biology and Molecular Epigenetics at the Faculty of Medicine, Department of Gastroenterology and Department of Experimental Medicine and Biotechnology, PGIMER, Chandigarh, India. His Ph.D. thesis concerned the investigation of the Wnt signaling pathway in search of new tumor prognostic and biomarkers in colorectal cancer. Dr. Anwar is the author of many book chapters and journal articles about tumor markers and is involved in other scientific activities in this advanced molecular technology era. He is a member of various scientific organizations and societies including the American Heart Association (AHA), American Society for Pharmacology and Experimental Therapeutics (ASPET), ASPET Young Scientists Committee, North American Vascular Biology Organization (NAVBO), United States, and Canadian Academy of Pathology (USCAP). He is the recipient of various awards including the Outstanding Young Investigator Travel Award and other postdoctoral competition category awards. He is also an editorial board member and reviewer for various scientific journals.

Dr. Zeenat Farooq is a postdoctoral research associate at the Department of Medicine, University of Illinois Chicago. Dr. Farooq obtained her Ph.D. from the School of Biotechnology (Chromatin and Epigenetics Lab), University of Kashmir, India. The focus of her Ph.D. was the elucidation of protein factors that play a role in the organization and maintenance of heterochromatin and regulation of transcriptional gene silencing. She has worked actively on various aspects of epigenetics and chromatin biology, including cancer biology, and has authored many papers in reputed journals. After completing her Ph.D., Dr. Farooq worked as a research associate at the Indian Council of Medical Research (ICMR), India, where her work focused on the role of epigenetic mechanisms involved in chronic complications of type 2 diabetes mellitus. She moved to the University of Illinois Chicago in 2020 and studied the role of regulation of mRNA translation under nutrient deprivation. Her recent work is focused on understanding the role of a novel hexokinase in the propagation of altered glucose metabolism and predisposition to Alzheimer's disease. Dr. Farooq has authored one book and many book chapters. She is a member of various scientific organizations and societies including the American Heart Association (AHA). In addition, she serves as a reviewer for various scientific journals. She is also a recipient of various awards and honors including a doctoral fellowship from CSIR-UGC, ASRB-NET, GATE, a Merit scholarship from the Department of Biotechnology (DBT), India, and an Independent Research Associateship from the Indian Council of Medical Research (ICMR).

Riyaz Ahmad Rather, Ph.D., is currently an assistant professor at the Department of Biotechnology, School of Natural and Computational Sciences, Wachemo University, Hossana, Ethiopia. Dr. Rather obtained his Ph.D. in Medical Biotechnology at VELS University, Chennai, India. His Ph.D. thesis concerned the elucidation of molecular mechanisms of action of various antioxidants as anti-cancer and anti-fungal agents in cancer genesis. He is the author of two books and has contributed many book chapters and journal articles on medical and molecular diagnostics. He is a member of various scientific organizations including the Biologists Forum of India, the Association of Basic Medical Scientists, and the European Atherosclerosis Society. Dr. Rather is the recipient of various awards including SGRF and EAS awards, among others.

Mohammad Tauseef, Ph.D., is currently an assistant professor at the Department of Pharmaceutical Sciences, College of Pharmacy, Chicago State University, Illinois, USA. Dr. Tauseef obtained his Ph.D. in Cardiovascular Pharmacology at the Department of Pharmacy, University of Delhi, New Delhi, India. His Ph.D. thesis concerned the investigation of inflammatory cell signaling molecules, with specific emphasis on vascular endothelial biology and cardiovascular pharmacology. He is the author of many book chapters and journal articles on molecular endothelial and cardiovascular biology and is involved in other scientific activities. He is a member of various scientific organizations and societies, including the American Heart Association (AHA). He is the recipient of various awards including an AHA postdoctoral fellowship, Midwest Affiliate, Outstanding Young Investigator Travel Award, Research Recognition award, Best Teacher award, and other postdoctoral competition category awards. He is also a reviewer for various scientific journals including the American Journal of Physiology-Lung Cellular and Molecular Physiology, Physiological Research, Journal of Cardiovascular Pharmacology and Therapeutics, Journal of Visualized Experiments, Plos One, and Human & Experimental Toxicology.

Thomas Heinbockel, Ph.D., is Professor and Interim Chair, Department of Anatomy, Howard University College of Medicine, Washington, DC. He holds an adjunct faculty position in both the Department of Anatomy and Neurobiology and the Department of Physiology, University of Maryland School of Medicine, Baltimore, MD. Dr. Heinbockel studied biology at Philipps-University Marburg, Germany. He began his studies of the brain during his MS thesis work at the Max Planck Institute for Behavioral Physiology, Starnberg/Seewiesen, Germany. Dr. Heinbockel earned a Ph.D. in Neuroscience at the University of Arizona. After graduating, he worked as a research associate at the Institute of Physiology, Otto-von-Guericke University, Magdeburg, Germany. Dr. Heinbockel's research is focused on understanding how the brain processes information as it relates to neurological and psychiatric disorders. His laboratory at Howard University concentrates on foundational and translational topics such as drug development, organization of the olfactory and limbic systems, and neural signaling and synaptic transmission in the central nervous system.

Contents

Preface

A few decades ago, research in biological sciences, especially molecular biology and disease connection, was confined to research on prominent diseases and involved techniques like polymerase chain reaction (PCR) to study gene polymorphisms, underpin the genetic nature and heritability of certain diseases, study mutations in various forms of cancer, and so on. These approaches, although still relevant and significant, were not able to fully capture the underlying problems in various diseases and missed the mark on the role of "environmental influences" in various diseases, which we now understand to play a paramount role in propagating or sometimes even driving various diseases. In addition, these environmental influences also impact our overall wellbeing and play a huge role in all aspects of life, including longevity, stress levels, and responses to stress since these takes into account various unappreciated factors like the impact of social influences, diet, nature of work, exposure level to chemicals, environment, ethnicity, geographical location, climatic conditions, and so on. Our environmental influences encompass anything and everything from where we are located to what we do and how we are exposed to various conditions. In addition to affecting our well-being in a "transient" manner, these may also influence our health, well-being, and the likelihood of disease conditions in a more stable and long-term manner. Part of the reason for this is that these influences not only affect signaling pathways to produce an immediate outcome, but they can also impact the expression of various genes in both the short and long term. The science of epigenetics deals with the study of environmental influences that have the power to alter gene-environment interactions and can result in stably inherited patterns of gene expression.

In Section 1 of this book, Chapter 1 sheds light on the meaning of epigenetics, its role, and the evolution of its definition.

Section 2 examines the various roles of epigenetics as well as advancements in the field. Chapter 2 examines the Evolution of the Epigenome as the blueprint for Carcinogenesis. Chapter 3 discusses the role of epigenetics, in particular DNA, in different aspects of diabetes. Chapter 4 presents the mechanism of diet–epigenome interactions and how dietary components could be used as "epidrugs" to reverse some epigenetic signatures for positive health outcomes in cancer prevention. Finally, Chapter 5 addresses the effect of social and environmental influences on health and wellbeing at the epigenetic level.

Research in neurobiology was plagued for the longest time by a lack of techniques to "turn on" or "turn off" essential genes and pathways in one cell or one part of the brain to pinpoint a function, a change in neuron firing, or change in a certain signaling pathway. The science of optogenetics, with its single-cell resolution, has made it possible to turn genes on and off in particular cells under the influence of light. This has allowed neurobiologists to improve the assessment of various neurological functions and disorders. Section 3 of this book provides some Background and Mechanisms Governing Optogenetics Chapter 6 provides information on the meaning of Cyanobacterial phytochromes in Optogenetics.

Chapter 7 discusses the Functional mechanism of proton pump-type rhodopsins found in various microorganisms as a potentially effective tool in optogenetics, and Chapter 8 discusses spatiotemporal regulation of cell-cell adhesions.

Embodying the latest research-based knowledge in epigenetics and optogenetics, this book fosters a deeper understanding of these two disciplines. It presents scientific data and information in an easy-to-follow manner to allow readers from various disciplines of biological sciences, especially undergraduate and graduate students, to develop a better understanding of epigenetics and optogenetics.

Mumtaz Anwar, Ph.D.
Research Assistant Professor at the Department of Pharmacology
and Regenerative Medicine,
College of Medicine,
University of Illinois at Chicago,
Illinois, USA

Dr. Zeenat Farooq
Postdoctoral Research Associate at the University of Illinois,
Chicago, USA

Riyaz Ahmad Rather, Ph.D.
Department of Biotechnology,
School of Natural and Computational Sciences,
Wachemo University,
Hossana, Ethiopia

Mohammad Tauseef, Ph.D.
Department of Pharmaceutical Sciences,
College of Pharmacy,
Chicago State University,
Illinois, USA

Thomas Heinbockel, Ph.D.
Professor and Interim Chair,
Department of Anatomy,
Howard University College of Medicine,
Washington, DC, USA

Section 1

Introduction to the Book

Chapter 1

Introductory Chapter: Epigenetics and Optogenetics - The Science behind the Cover Blanket of Our Genome

Mumtaz Anwar, Thomas Heinbockel and Zeenat Farooq

1. Introduction

1.1 Epigenetics

For the longest time in the history of scientific research, a belief existed that DNA, the master molecule that makes up our genome, is the destination of living beings, the blueprint for every trait and disease that we might inherit or develop. Various landmark discoveries through many decades contributed to this "ultimate destination" tag of the DNA like the double helical structure in 1953 by Watson and crick, discovery of mutations in certain genes contributing to disease phenotypes such as phenylketonuria, cystic fibrosis, p53, and many more. These developments led to immense interest in the field and one of the most astounding accomplishments in this regard was the "human genome project," which resulted in complete sequencing of the human genome. Soon after, complete genome sequences of closely related organisms and other model organisms were deciphered, published, and made available for use by every researcher across the globe. This led to the inception of the fields of bioinformatics and comparative genetics.

In the middle of all the euphoria about research on DNA and genes, it was being increasingly realized that only about 2% of the DNA in humans codes for proteins. The rest of the DNA was initially called junk DNA. However, an intriguing question surfaced regarding the reason for nature to preserve this huge amount (98%) of junk DNA if it did not serve any function. This seemed quite paradoxical to the concept of evolution.

This question paved the way for more research, and soon interest started booming in the field of epigenetics. This term has been used differently by different scientists from time to time, according to what could be proven using the resources and technology of that time. Epigenetics (from epigenesis) was first aimed to describe changes that take place when a zygote undergoes divisions and leads to differentiation (genesis) into different cell types, tissues, and organs. It was a beautiful concept to illustrate the differentiation potential of zygote, but the knowledge of the mechanisms responsible for this potential was lacking at that time. The term was originally coined by C.H. Waddington in 1942 as the phenomenon that changes the cells from totipotent state to fully differentiated state during embryonic development [1].

The phenomenon of heredity and the concept of genes were not known back then, and hence these definitions did not contain any molecular feature. Later, the term was defined by Riggs as "the study of mitotically and/or meiotically heritable changes in gene function that cannot be explained by changes in DNA sequence" [2]. The most common definition of epigenetics today is "the study of phenomena that lead to heritable changes in gene expression without changing the sequence of nucleotides." For the sake of simplicity and universality, an epigenetic trait was defined as "a stably heritable phenotype resulting from changes in a chromosome without alterations in the DNA sequence" at the Cold Spring Harbor meeting in 2008 [3].

All these definitions were based on two important principles.

I. The change should influence gene expression and not DNA itself.

II. The change should be heritable.

With the discovery of histones, it was initially thought that these proteins only helped the DNA to wrap itself appropriately to fit into the nucleus. However, with advancing research, histones were viewed as the "interface" between DNA and the environment. These were the proteins that could change the accessibility of genes within the DNA to increase or decrease expression and interestingly, they could do it without the requirement to change the sequence of the underlying gene. This led to the identification of various histone modifications such as methylation, acetylation, phosphorylation, ubiquitylation, and so on, each one of them having their own kind of impact, that is, either increasing or decreasing gene expression. Research performed in the field also showed that different cells carry different combinations of histone modifications, and these combinations together constitute the *histone code*. More research on histones identified mechanisms such as histone sliding that can also influence gene accessibility and expression in response to various signaling pathways and at different stages of the cell cycle. DNA methylation on the 5' cytosine also came to be recognized as a mechanism that could impact gene expression independent of the sequence of the gene that carries them.

Further research on model organisms was conducted on histones and DNA methylation to establish the transmissibility of epigenetic traits at the molecular level [4, 5]. One of the pioneering experiments performed on the mechanism of epigenetic inheritance was carried out by Manel Esteller and colleagues. The group extensively studied identical twins and verified that twin pairs that were older and/or had experienced different lifestyles had far greater differences in epigenetic marks (histone acetylation and DNA methylation) [6]. These studies were astounding as they helped in establishing the fact that DNA is not the destination to dictate all traits but patterns of expression and epigenetic changes can result in the establishment of different traits as a result of different environments, even in identical twins. Another study showed that supplementation of the diet of expectant mice with vitamin B, folic acid, choline, and betaine could alter the color of the fur of their offspring by affecting DNA methylation of the pigmentation genes [7]. Research on the agouti gene, which can cause diabetes and yellow color pigmentation of the fur in mice, has shown that offspring born to mice that were fed with supplements that resulted in methylation of the gene were slim and nondiabetic due to increased DNA methylation and consequent silencing of the agouti gene [8]. These experiments proved beyond doubt that we not only inherit our parents' DNA but also their experiences and exposures, which influence our traits. Studies performed concomitantly and afterward also showed how the exposure of mothers to conditions such as smoking, alcohol consumption, stress during pregnancy, prenatal malnutrition, etc., can influence epigenetic patterns of key genes in offspring [9].

The very fact that epigenetic changes are heritable, yet reversible stimulated a lot of interest in the field because it provided a ray of hope to find a cure for many diseases that were initially thought to be terminal. This effect also impacts directly at the level of gene expression and hence can offer a lasting and more effective therapeutic approaches [10]. In addition, it established that different cells carry different epigenetic signatures, and that one cell type can be changed to another, or a *diseased* cell can be converted into a *healthy* one through changes in the epigenetic landscape.

More research identified more players of the field such as long non-coding RNAs, enhancer RNAs, micro RNAs, etc. It was in fact realized that the so-called "junk DNA" actually codes for these "regulator elements," which play a role in regulating the expression of the genes that code for proteins [11]. Until now, we have been able to decipher very little information about epigenetic or regulatory elements. The fact that 98% of the genome codes for regulatory elements prompt us to believe that the field of epigenetics is very diverse and yet mostly unexplored. If this field is explored with the help of more advanced research tools and technology in the future, we might be able to find cures for many debilitating diseases of humans, might find more answers for our similarities and dissimilarities with other species, better understand evolution, and might develop a better understanding of the entire ecosystem by unraveling more connections related to gene–environment mechanisms. Increased knowledge of how gene–environment interactions operate acquired by means of increased knowledge of epigenetics through superior technology might answer many ecologically important questions for us and might enable us to understand the ecosystem and the role of *Homo sapiens* in this ecosystem in relation to other species and the environment more clearly and effectively.

2. Optogenetics

The sequencing of the genome in species as different as humans and plants has helped us to understand mechanisms of development, physiology, and evolution [12–14]. The field of epigenetics studies chemical modifications of the DNA as well as interactions that include genome-associated proteins to analyze differences in the expression of genes that are heritable and arise without a change of the DNA sequence. As such epigenetic mechanisms afford another mechanism of transcriptional control in regulating gene expression. While the field of epigenetics revealed an entire new layer of genetic regulation, optogenetics is the field that has allowed researchers to study cell signaling pathways and networks with unprecedented detail and resolution [15, 16]. This relatively new field exemplifies the power of taking a molecular approach to explore complex biological systems such as the brain in order to understand even the nature of emotions or psychiatric disorders [17]. Optogenetics is a combination of genetic manipulation and the use of optical tools. Genes that confer light responsiveness are inserted into cells of interest and allow for subsequent assessment of well-defined events in cells or even freely moving animals. Genetic tools allow the insertion of genes into cells that afterward respond to specific wavelengths of light. Subsequently, light can turn on or off specific signal cascades in cells and even trigger or inhibit the behavior of organisms. Thereby, optogenetics gives researchers an opportunity to obtain a deep view into an organism under optical control [18].

To understand the brain means to be able to reliably manipulate it and predict its response. Neuroscientists have long used electrophysiological techniques to stimulate particular brain areas or even single neurons [15]. Electrical stimuli activate neural circuitry, often without being able to stop neuronal activity. Neuropharmacological tools are based on drugs that are slow in their effects or not specific enough to stimulate individual cells. In 2005, a set of new techniques started to emerge that combined optical stimuli with genetic tools in order to control events in individual cells [19]. The field of optogenetics has since revolutionized experimental approaches to study cell signaling, metabolism, brain circuits, and organismal behavior.

Two pieces of information about the origin of the field are worth mentioning. As recounted by Karl Deisseroth [15], it was Nobel Laureate Francis Crick who suggested the creation of this new field in the late 1970s by stating that the major challenge facing neuroscience was the need to control one type of brain cell while leaving others unaltered. Later on, Crick proposed the use of light to achieve this control feat because it could be delivered in precisely timed pulses. The other piece of information relates to the fact that it was microorganisms that allowed optogenetics to come into existence. It had been known for many years that certain microorganisms generate proteins, which allow ions to cross the cell membrane in response to light. The genes coding for these proteins are known as opsins. One of the proteins, bacteriorhodopsin, discovered in 1971, is an ion pump that can be activated by photons of green light [20]. Later on, other opsins were identified, namely the halorhodopsins and channel rhodopsins, which are also light-gated ion pumps, more specifically, single-component light-activated cation channels. These discoveries have led to widespread use of optogenetic tools. Channelrhodopsin-1 (ChR1) and Channelrhodopsin-2 (ChR2) are found in the model organism *Chlamydomonas reinhardtii*. In 2005, several groups published the first accounts of using ChR2 as a tool for genetically targeted optical remote control, namely optogenetics, of neurons, neural circuits, and behavior of animals [19, 21, 22]. This marked the beginning of the field of optogenetics. Optogenetics has taken advantage of microbial opsins such as channel rhodopsin to genetically target and then remotely control excitable cells. In order to control cells or organisms, optical activation is superior to other methods because of its speed, ease of use, specific targeting, and precise temporal control of optical activation.

Acknowledgements

This publication resulted in part from research support to T.H. from the National Science Foundation [NSF IOS-1355034], Howard University College of Medicine, and the District of Columbia Center for AIDS Research, an NIH funded program [P30AI117970], which is supported by the following NIH Co-Funding and Participating Institutes and Centers: NIAID, NCI, NICHD, NHLBI, NIDA, NIMH, NIA, NIDDK, NIMHD, NIDCR, NINR, FIC, and OAR. The content is solely the responsibility of the authors and does not necessarily represent the official views of the NIH.

Conflict of interest

The authors declare that there is no conflict of interest regarding the publication of this chapter.

Author details

Mumtaz Anwar[1]*, Thomas Heinbockel[2] and Zeenat Farooq[1]

1 Department of Pharmacology and Regenerative Medicine, College of Medicine, University of Illinois at Chicago, Chicago, USA

2 Department of Anatomy, College of Medicine, Howard University, Washington, DC, USA

*Address all correspondence to: mumtazan@uic.edu;
mumtaz_anwar1985@yahoo.co.in

IntechOpen

References

[1] Waddington CH. The Epigenetics of Birds. (Verlag) 294 Seiten: Cambridge University Press; 2014. ISBN: 978-1-107-44047-0

[2] Riggs AD, Martienssen RA, Russo VE. *Epigenetic Mechanisms of Gene Regulation*. Plainview, NY: Cold Spring Harbor Laboratory Press; 1996. pp. 1-4. ISBN 978-0-87969-490-6

[3] Berger SL, Kouzarides T, Shiekhattar R, Shilatifard A. An operational definition of epigenetics. Genes & Development. 2009;**23**(7): 781-783

[4] Gayon J. From Mendel to epigenetics: History of genetics. Comptes Rendus Biologies. 2016;**339**(7-8):225-230

[5] Skvortsova K, Iovino N, Bogdanović O. Functions and mechanisms of epigenetic inheritance in animals. Nature Reviews Molecular Cell Biology. 2018;**19**(12):774-790

[6] Fraga MF, Ballestar E, Paz MF, Ropero S, Setien F, Ballestar ML, et al. Epigenetic differences arise during the lifetime of monozygotic twins. Proceedings of the National Academy of Sciences of the United States of America. 2005;**102**(30):10604-10609

[7] Zeisel S. Choline, other methyl-donors and epigenetics. Nutrients. 2017;**9**(5):445

[8] Dolinoy DC. The agouti mouse model: An epigenetic biosensor for nutritional and environmental alterations on the fetal epigenome. Nutrition Reviews. 2008;**66**(Suppl 1):S7-S11

[9] Ooi SL, Henikoff S. Germline histone dynamics and epigenetics. Current Opinion in Cell Biology. 2007;**19**(3): 257-265

[10] Szyf M. Prospects for medications to reverse causative epigenetic processes in neuropsychiatry disorders. Neuropsychopharmacology. 2017;**42**(1): 367-368

[11] Wei JW, Huang K, Yang C, Kang CS. Non-coding RNAs as regulators in epigenetics (Review). Oncology Reports. 2017;**37**(1):3-9

[12] International Human Genome Sequencing Consortium. Initial sequencing and analysis of the human genome. Nature. 2001;**409**:860-921

[13] Venter JC, Adams MD, Myers EW, Li PW, Mural RJ, Sutton GG, et al. The sequence of the human genome. Science. 2001;**291**:1304-1351

[14] Green JD, Watson JD, Collins FS. Human genome project: Twenty-five years of big biology. Nature. 2015;**526**: 29-31

[15] Deisseroth K. Optogenetics: 10 years of microbial opsins in neuroscience. Nature Neuroscience. 2015;**18**:1213-1225

[16] Zhou X, Mehta S, Zhang J. Genetically encodable fluorescent and bioluminescent biosensors light up signaling networks. Trends in Biochemical Sciences. 2020;**45**(10):889-905. DOI: 10.1016/j.tibs.2020.06.001 Epub 2020 Jul 10. PMID: 32660810; PMCID: PMC7502535

[17] Deisseroth K. From microbial membrane proteins to the mysteries of emotion. Cell. 2021:S0092-8674(21)00992-2. DOI: 10.1016/j.cell.2021.08.018 Epub ahead of print. PMID: 34562367

[18] Greenwald EC, Mehta S, Zhang J. Genetically encoded fluorescent biosensors illuminate the spatiotemporal regulation of signaling networks. Chemical Reviews. 2018;**118**(24):11707-11794. DOI: 10.1021/acs.chemrev.8b00333 Epub 2018 Dec 14. PMID: 30550275; PMCID: PMC7462118

[19] Boyden ES, Zhang F, Bamberg E, Nagel G, Deisseroth K. Millisecond-timescale, genetically targeted optical control of neural activity. Nature Neuroscience. 2005;**8**(9):1263-1268. DOI: 10.1038/nn1525 Epub 2005 Aug 14. PMID: 16116447

[20] Oesterhelt D, Stoeckenius W. Rhodopsin-like protein from the purple membrane of Halobacterium halobium. Nature: New Biology. 1971;**233**(39): 149-152. DOI: 10.1038/newbio233149a0 PMID: 4940442

[21] Li X, Gutierrez DV, Hanson MG, Han J, Mark MD, Chiel H, et al. Fast noninvasive activation and inhibition of neural and network activity by vertebrate rhodopsin and green algae channelrhodopsin. Proceedings of the National Academy of Sciences of the United States of America. 2005; **102**(49):17816-17821. DOI: 10.1073/ pnas.0509030102 Epub 2005 Nov 23. PMID: 16306259; PMCID: PMC1292990

[22] Nagel G, Brauner M, Liewald JF, Adeishvili N, Bamberg E, Gottschalk A. Light activation of channelrhodopsin-2 in excitable cells of Caenorhabditis elegans triggers rapid behavioral responses. Current Biology. 2005; **15**(24):2279-2284. DOI: 10.1016/j. cub.2005.11.032 PMID: 16360690

Section 2

Different Facets of Epigenetics

Chapter 2

Evolution of Epigenome as the Blueprint for Carcinogenesis

Zeenat Farooq, Ambreen Shah, Mohammad Tauseef,
Riyaz Ahmad Rather and Mumtaz Anwar

Abstract

Epigenetics "above or over genetics" is the term used for processes that result in modifications which are stably inherited through cell generations, without changing the underlying DNA sequence of the cell. These include DNA methylation, Post-translational histone modification and non-coding RNAs. Over the last two decades, interest in the field of epigenetics has grown manifold because of the realization of its involvement in key cellular and pathological processes beyond what was initially anticipated. Epigenetics and chromatin biology have been underscored to play key roles in diseases like cancer. The landscape of different epigenetic signatures can vary considerably from one cancer type to another, and even from one ethnic group to another in the case of same cancer. This chapter discusses the emerging role of epigenetics and chromatin biology in the field of cancer research. It discusses about the different forms of epigenetic mechanisms and their respective role in carcinogenesis in the light of emerging research.

Keywords: Epigenetics, DNA Methylation, Histone Modifications, Cancer

1. Introduction

Transmission of characters in a stable, inheritable manner is governed by the genetic make-up of a cell. This information for vertical transmission of characters is carried by the macromolecule deoxyribonucleic acid (DNA). The linear sequence of nucleotides in the DNA dictates the sequence of amino acids in the proteins and hence controls all the vital processes occurring within the cell. However, the linear length of DNA molecules is very long. For example, a typical human cell contains about 2 meters long DNA. Therefore, in order to accommodate DNA into nucleus, this genetic information is contained in the form of a nucleoprotein complex called chromatin [1]. This is particularly true about eukaryotic cells. Though prokaryotic cells also contain a nucleoid, it, however, is not well-organized.

The organization of DNA into chromatin is particularly important for two main reasons.

1. To bring about compaction of the large DNA molecule into a small nuclear space in an ordered manner.

2. To facilitate regulated gene expression.

Alongside DNA, chromatin mainly consists of small, basic positively charged group of proteins called histones. The positively charged histones bind with the negatively charged DNA in an energetically favorable manner inside chromatin [2]. These proteins have remained the focus of intensive research for many years now. Apart from DNA and histones, chromatin also contains a huge array of non-histone proteins, most of which are not as well characterized and well-studied as histones.

Earlier it was thought that compaction of DNA into chromatin solely occurs to accommodate DNA. But later it was realized that this compaction plays a paramount role in orderly organization of DNA and thereby helps in differential gene expression. The fundamental repeating unit of chromatin is the nucleosome which consists of two copies each of histones H2A, H2B, H3 and H4 wrapped around 146 bp of DNA in a left-handed helical manner [1]. The histone proteins are named in the order in which they were discovered. Because of being associated with the nucleosome core, these histone proteins are known as the core histones. Another class of histones binds DNA at the entry and exit sites into nucleosomes. This is known as the linker histone H1 and paves way for further compaction of nucleosomes into higher order chromatin structures (**Figure 1**).

Upon observation under a microscope, chromatin appears as two distinct entities within the nucleus. These are termed as euchromatin and heterochromatin. Euchromatin is the lightly stained part of chromatin which mostly lies towards the interior regions of nucleus and contains actively transcribed genomic regions. Heterochromatin is the darkly stained fraction which mostly lies towards the periphery of nucleus [3]. It contains regions which are transcriptionally silent and mostly contains repetitive DNA sequences. This spatial organization of chromatin is maintained through various mechanisms. These mechanisms serve as the "epigenetic carriers of nuclear information" within the cell and include covalent histone modifications, non-coding RNAs and chromatin remodeling complexes and lately also included DNA methylation (**Figure 2**).

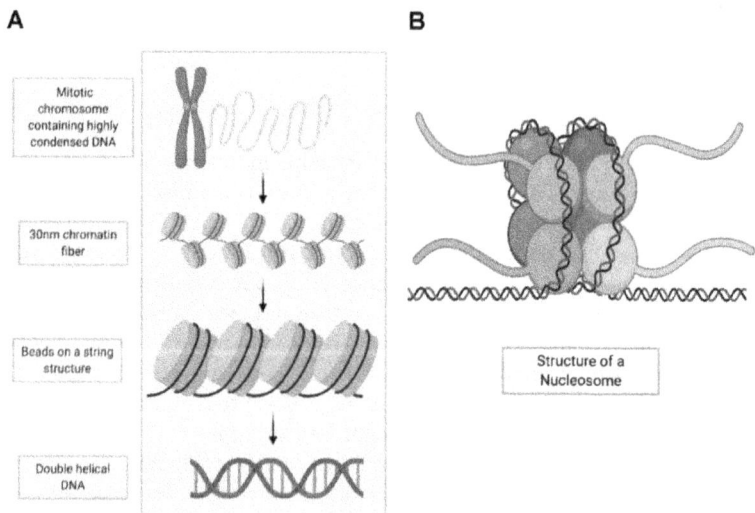

Figure 1.
Representation of different levels of hierarchical chromatin organization. (A) Inside a compact chromosome, DNA and proteins are organized at different levels. (B) Ultrastructure of a nucleosome containing two copies of H2A,H2B,H3 and H4 inside 147 bp of DNA.

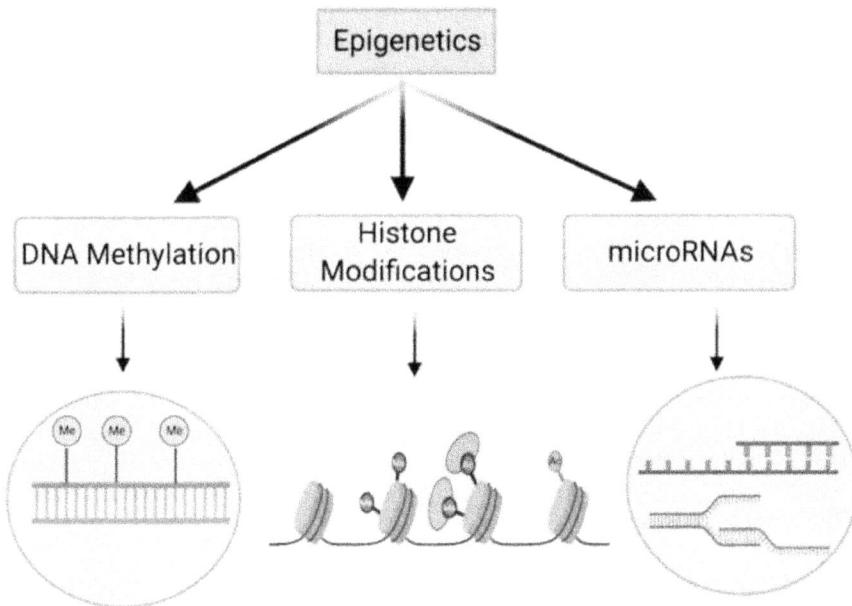

Figure 2.
Major players involved in the propagation of epigenetic mechanisms in cells. DNA methylation and micro RNAs are involved in gene silencing, histone modifications are involved in both silencing and expression of genes.

2. Epigenetics and chromatin biology: unifying themes and differences

At its heart, epigenetics refers to the study of heritable changes in gene expression without changes in the DNA sequence. This term was coined by Waddington and as the name indicates, epi (above or over genetics) is any moiety that can be stably inherited by cells across many generations without altering the sequence of nucleotides in the DNA. The study of epigenetics previously involved study of covalent histone modifications and non-coding RNAs. However, DNA methylation has also been increasingly recognized as an epigenetic phenomenon owing to its non-sequence based heritable nature and its importance in maintaining cellular homeostasis and association of its perturbations with various diseases. Therefore, the definition and scope of epigenetics has changed dynamically since the inception of the field.

Quite often, epigenetics and chromatin biology are very loosely stated terms. However, to be more precise, epigenetics refers to the study of "epigenetic marks or signatures" which play a prominent role in maintenance of cellular homeostasis whereas chromatin biology refers to the study of "chromatin structure and function". This encompasses nuclear dynamics, topology, localisation, organisation and three-dimensional (3D) structure [3]. There is a huge overlap between the two terms, and these are often used interchangeably. For example, epigenetic signatures and modifications play a paramount role in the maintenance of nuclear topology, overall chromatin organization and chromatin states.

Field of epigenetics is very interesting because of the reversible nature of epigenetic changes. This means that although these changes can be stably inherited, however, unlike DNA sequence, these changes can also be reversed under particular conditions. In fact, mechanisms are well in place within the cells which lead to the reversal of these modifications [4]. Interestingly, these changes can also be targeted for the reversal externally, using specific enzymes, under desired conditions. This may include the reversal of epigenetic modifications involved in disease progression with the help of enzymes [5]. For example, reversal of an epigenetic modification that is involved in carcinogenesis by an enzyme specific for the reversal to alleviate some of the symptoms.

Epigenetic modifications play a very prominent role in almost all the cellular processes like growth, cell division, maintenance of cellular identity etc. Therefore, any changes in these modifications can lead to serious outcomes. Perturbations in epigenetic modifications have been observed to be involved in various deleterious conditions including cancer [6].

In this chapter, we shall discuss about the various epigenetic mechanisms, their importance, major functions that they carry out in the cells and changes to these marks and their implications in cancer.

3. DNA methylation

DNA methylation involves transfer of a methyl group from S-adenosylmethionine to the 5'position of cytosine residues in DNA. DNA methylation is one of the most prominent epigenetic events that take place within the cells and has been shown to play important roles in various cellular processes like genome integrity, genome imprinting, X chromosome inactivation and development [7–9].

DNA methylation at 5 methyl cytosine is catalyzed by two groups of methyltransferases.

1. DNMT1 which catalyzes methylation on the newly synthesized hemi-methylated DNA strand, utilizing the parental strand as template for copying of methylation pattern. This class of enzymes are known as the maintenance methyltransferases as they play role in maintaining the methylation status following replication. These are critically important enzymes for mammals as mice deficient in DMNT1 display embryonic lethality [10].

2. DNMT3a and 3b. These are the enzymes which play role in methylating DNA at 5' methyl cytosine without utilization of a methylated template. These enzymes are therefore known as *de novo* methyltransferases and these have been known to catalyze methylation events during various important cellular phases like development. These enzymes are therefore highly expressed during embryogenesis and display reduction in expression pattern in adult tissues [11]. DNMT 3a and 3b are also extremely important for mammals since DNMT 3b deficient mice, similar to DNMT 1, are embryonic lethal whereas those deficient in DNMT 3a die by the age of 4 weeks [10].

Another member of the DNMT family of enzymes is DNMT 3 L. It was discovered in 2000. DNMT 3 L lacks an intrinsic methyltransferase activity but assists DNMNT3a and 3b in methylating retrotransposons [12].

In eukaryotes, DNA methylation occurs predominantly within repetitive sequences in order to maintain genomic integrity [13]. Methylation on cytosine residues usually takes place in the context of CG dinucleotides (Known as CpG) and around 75% of CpG dinucleotides in humans remain methylated. These CpG dinucleotides are unevenly distributed but are concentrated in stretches of high frequency known as CpG islands. These islands remain mostly unmethylated and can be found in the promoters of constitutively expressed genes like housekeeping genes [14]. In humans, almost half of the estimated 29,000 CpG islands remains unmethylated under normal conditions [15–17].

Methycytosine residues often co-operate with other effectors to bring about a silenced chromatin state. Methyl binding domain (MBD) proteins recognize and bind to methylated cytosines. These MBD proteins act as a signal/binding platform for histone modifying and chromatin remodeling enzymes to bring about further compaction of chromatin [18]. Apart from binding methylated DNA, MBD 2 (a member of MBD family of proteins) has also been shown to promote the DNA methyltransferase activity of NuRD (chromatin remodeling complex) by interacting with NuRD [19, 20]. This interaction brings NuRD complex in close proximity of cytosine residues which are later methylated by NuRD. Till date, six members of methyl binding domain proteins have been identified that include MBD1, MBD2, MBD3, MBD4, methylcytosine binding protein 2 (MECP2) and Kaiso [21]. All of these proteins are under intense investigation and efforts are being made to identify more members of the family.

Various genes contain regions of CpG dinucleotides in their promoters with variable degrees of methylation levels [14]. These levels are crucial for normal functioning of the cells and any mis-regulation in this level is associated with a number of physiological outcomes. Methylated DNA elements often co-operate with other epigenetic elements to ensure proper silencing of chromatin and any increase in levels of DNA methylation are often involved in silencing of cognate genes which can lead to carcinogenesis [15, 22]. For example, it has been observed that increase in the levels of promoter DNA methylation in tumor suppressor genes leads to a decrease in their expression and hence a steady decline in their cellular activity is observed [15, 23–25]. Hypermethylated promoters can also serve as targets for transition mutations due to spontaneous deamination of 5'methyl cytosine into thymine [7, 26]. This leads to transmission of DNA with errors during replication to new cells. These cells are genomically unstable and with time, accumulate more and more mutations which in the absence of proper surveillance, eventually lead to cancer initiation [7, 16, 27]. Decrease in the DNA methylation of tumor suppressor genes has been observed in a number of primary tissues from cancer patients at various geographical locations.

Global hypomethylation can also ensue which can lead to loss of repression from the repetitive DNA sequences (like transposons) and imprinted genomic sequences. This can be accompanied by loss of methylation from genomic regions involved in maintaining chromosome stability like peri centromere. This can cause gross genomic instability which is a characteristic of many forms of cancer. Though the relationship between global loss of DNA methylation and cancer has not been very well studied and needs more research (**Figure 3**) [16, 28, 29].

Alternatively, certain genes undergo hypomethylation and therefore experience increase in expression that has been associated with carcinogenesis. Genes predominantly affected by hypomethylation include developmentally critical genes, enzymes, growth regulatory genes and tissue-specific genes such as germ cell-specific tumour antigen genes [30]. Various other genes which have been shown to be involved in carcinogenesis as a result of aberrant DNA methylation are listed in **Table 1**.

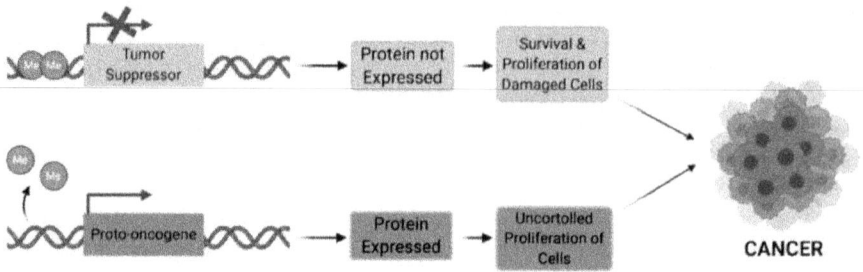

Figure 3.
Schematic of two broad mechanisms involved in cancer progression through DNA methylation. Hypermethylation and silencing of tumor suppressor gene promoters to allow unchecked growth of damaged cells to accumulate more damage and generate cancer phenotype. Hypomethylation of proto-oncogenes to favor uncontrolled proliferation of cells to generate cancer mass.

S.No.	Name of gene	DNA methylation change	Change in gene expression	Type of cancer	References
1.	P16	Increase	Decrease	Colorectal, Renal Lung, Oral, Head and neck, Hepatic	[23, 29, 31–39]
2.	Hmlh1 and hMSH2	Increase	Decrease	Colorectal, Renal	[40–42]
3.	P Cadherin	Increase	Decrease	Breast, Hepatic, Pancreatic, Lung, Salivary gland	[26, 37, 43–45]
4.	Cyclin D2	Decrease		Gastric	[46]
5.	MAGE	Decrease		Melanoma	[47]
6.	P15	Increase	Decrease	Oral carcinoma	[32]
7.	RASSF1	Increase	Decrease	Nasopharyngeal Hepatic, Bladder	[37, 48–50]
8.	MGMT	Increase	Decrease	Oral, Head and neck Bladder, Lung	[33, 35, 38, 39, 50, 51]
9.	FHIT	Increase	Decrease	Lung	[23, 43]
10.	DAP-K	Increase	Decrease	Oral, Nasopharyngeal Head and neck, Lung Pancreatic, Renal	[32, 33, 35, 48, 38, 39, 51–53]
11.	APC	Increase	Decrease	Colorectal, Lung	[40, 51, 54]
12.	RAR (retinoic acid receptor)			Nasopharyngeal Head and Neck Lung	[23, 38, 39, 43]

Table 1.
Changes in DNA methylation of different genes in different forms of cancer.

4. Epigenetic modifications in context of chromatin

The organization of DNA into chromatin, although very necessary, imposes constraints on all the nuclear processes which require DNA as a template like replication, transcription and repair. Therefore, in order to gain access to the underlying DNA, chromatin structure is dynamically regulated through various mechanisms. This flexibility is permitted by mechanisms like histone modifications, incorporation of histone variants and chromatin remodeling [2].

Histone modifications act as binding platforms for various effectors for appropriate downstream signaling. Histone variants are incorporated by replacing canonical histones under specified conditions into nucleosomes. The variants possess different bio-physical properties compared to their canonical counterparts and hence play crucial roles in cellular processes like DNA repair. Chromatin remodeling leads to sliding of nucleosomes along chromatin, exposing regions of genome which could be acted upon by trans-acting factors for specified outcomes.

4.1 Histone modifications

Histone proteins undergo a variety of covalent modifications which can either lead to compaction or relaxation of the underlying DNA within chromatin. The outcome of these modifications is dictated by the type of modification, degree of modification as well as stage of the cell cycle. Histone proteins consist of a highly structured C-terminal globular domain and an unstructured N-terminal tail. Globular domains are generally involved in mediating histone-histone and histone-DNA interactions while as N-terminal tails act as sites for covalent modifications. Among the different classes of histone proteins, histone H3 and H4 generally undergo covalent modifications in their tails. Though recently, H2A and H2B have also been observed to undergo certain modifications [55, 56]. Similarly, many modifications have been observed in the globular domain of histone H3 as well [1]. Histone modifications play role in numerous biological processes like gene regulation, DNA repair, chromosome condensation and spermatogenesis [57]. Some of the well-recognized histone modifications include acetylation and ubiquitination of lysine (K) residues, phosphorylation of serine (S) and threonine (T) residues, methylation of arginine (R) and lysine (K) residues as well as other less known modifications [58, 59]. These modifications are largely postulated to affect chromatin function through two distinct mechanisms: By altering the electrostatic charge of histones, these could alter the structural properties or the binding of histones to DNA. As against the first mechanism, some of the modifications create binding surfaces for the recruitment of specific functional complexes to their sites of action e.g., proteins containing bromodomains recognize acetylated residues while those containing chromodomains recognize methylated residues [60, 61]. It was, In fact, the potential specificity of these interactions which prompted Struhl and Allis to propose the '*histone code hypothesis*' according to which "*specific combinatorial sets of histone modification signals dictate the recruitment of particular trans-acting factors to accomplish specific functions*" [62]. Initially, it was thought that histone proteins undergo covalent modifications after translation (post translational modifications) in a manner dictated by nucleosomal context. But recently, it has been observed that histones can undergo co-translational modifications as well, depending upon the cellular context. This observation has added an additional layer into the role of histones in regulation of cellular homeostasis and clearly calls for more research in the field. Perturbations in histone modifications is associated with many physiological disturbances, including carcinogenesis [5].

4.1.1 Histone acetylation and deacetylation

Acetylation is the most widely studied post translational modification in histones. This modification involves transfer of an acetyl group from N-acetyl-Co-A to the € amino group of lysine with the help of histone acetyltransferases (HATs). Histone acetylation is associated with loosing of chromatin structure due to neutralization of the positive charge on histones with the negative charge on acetyl group which is responsible for increase in transcription. In fact, various transcription activator or co-activator complexes contain HAT activity such as CBP 300, TAF II 250. Reversal of acetylation is carried out by another class of enzymes known as histone deactylases (HDACs). Both HATs and HDACs have been studied extensively in relation to various diseases, including neurodegeneration and cancer [4]. Depending upon the gene/s being involved (oncogenes or tumor suppressor genes), HATs and HDACs can have different effects on the cancer outcomes.

4.1.2 HATs, HDACs and cancer

Relationship between histone acetylation status and cancer has been demonstrated in various studies. For example, a loss of acetylation on lysine 16 of histone H4 (H4K16) has been observed in cancer cell lines and primary human tissues by Fraga et al. [63]. Decrease in promoter acetylation and consequent decline in expression of P21 gene has been observed in some forms of cancer with subsequent rescue of expression upon treatment of cells with HDAC inhibitors under similar conditions [64]. Another study has linked decrease in histone acetylation with tumor invasiveness and metastasis [65]. Accumulating data also shows that HDACs are involved in hematological malignancies like acute promyelocytic leukemia (APL) due to aberrant recruitment to non-target promoters, as a result of interaction with translocation-induced fusion proteins like RAR-PML [66]. Downregulation of E-cadherin due to decrease in promoter acetylation levels has been implicated in the invasive potential by carcinomas [67, 68]. A number of studies have also linked levels of specific classes of HDAC enzymes with different forms of cancer like increase in HDAC1 expression in gastric [69], prostate [70], colon [71], breast carcinoma [72], increase in HDAC2 expression in cervical [73], gastric [74] and colorectal carcinoma [75]; increase in HDAC3 expression in colon carcinoma [76] and increase in HDAC 6 in breast carcinoma [71]. Mutations in HDAC2 gene has also been reported in sporadic colorectal carcinomas [77].

Various mechanisms are responsible for the role of specific forms of enzymes in specific cancer types, largely depending upon their interaction partners and the pathways involved. For example, HDAC1 has been shown to play a role in transcriptional repression of various oncogenic targets of retinoblastoma gene (Rb). Therefore, loss of HDAC1 activity leads to compromise in efficiency of Rb in downregulation of target oncogenes [78]. HDAC3 has also been seen to interact with retinoblastoma protein (Prb) in cancer, Perhaps the most important HDAC III enzyme in cancer is SIRT1 due to its role in regulation of protein factors like P53 [79], androgen receptor [80], p300 [81], E2F1 [82], DNA repair factor ku70 [83] and most importantly, NF-KB [84].

4.1.3 Histone methylation

Histone methylation involves transfer of methyl group(s) from S-adenosyl-methionine to lysine or arginine residues on histones. The enzymes catalyzing histone methylation are known as histone methyltransferases (HMTs). Depending on the target residue, histone methyltransferases are of two kinds 1. Histone lysine methyltransferases (HKMTs) and histone arginine methyltransferases (HRMTs). Also, lysine residues have three replaceable amino groups on the β-carbon. Therefore, lysine can undergo mono, di or tri-methylation whereas arginine can undergo only mon and di methylation.

Histone methylation is most commonly observed on lysine residues of H3 and H4 tails [85]. It is the most diverse histone modification in terms of complexity and is involved in various functions, depending on the physiological context. Histone methylations commonly associated with gene activation include H3K4, H3K36 and H3K79 and those associated with gene inactivation include H3K9, H3K27 and H4K20 [86]. Furthermore, variations in the degree of methylation on a single residue can also amplify the histone code further. For example, monomethylated H4K20 (H4K20me1) is involved in the compaction of chromatin and therefore transcriptional repression. However, H4K20me2 is associated with repair of DNA damage [63].

Histone methylation is involved in several cellular functions like maintenance of chromatin structure, DNA repair, gene silencing, prevention of hyper-recombination, maintenance of genome integrity et cetera. It is also involved in maintenance of X-chromosome integrity and silencing through excessively methylation of H3K9 on the second copy of human X chromosome in female cells. This provides a binding surface for methyl domain binding (MDB) protein and heterochromatin protein (HP1) to heterochromatinize and silence the second copy of X-chromosome [87, 88]. Since histone methylation plays a paramount role in regulation of gene expression and represents the most stable and complex histone modification, even slight changes to the methylation pattern can have deleterious effects on the organism. In *Saccharomyces cerevisiae*, a lethal mutation that leads to H3K4, H3K36 and H3K79 methylation inactivates many genes required for cell cycle progression and hence causes a delay in mitosis. It has been discovered that deletion of the methyltransferase genes which play role in the above-mentioned methylations allows this organism to live since the lysine residues in question are not methylated [89].

4.1.4 HMTs and cancer

Cancer cells use a diverse range of molecular mechanisms to alter histone methylation landscape. These include mis-regulation of histone methyltransferases and/or demethylases, mistargeting of histone methyltransferases and mutations in methyltransferases. For example, if areas around oncogenes become unmethylated, these genes will attain the potential of being transcribed at an alarming rate. On the contrary, if areas around tumor suppressor genes become highly methylated, these genes will lose their activity and therefore cancer will be more likely to occur [90]. Accumulating data suggests that histone methylation is mis-regulated in various forms of cancer [91, 92]. Fraga et al. [63] have observed that loss of H4K20 trimethylation that leads to hypomethylation of repetitive sequences is a common event in human cancers which occurs at a early stage during tumorigenesis. Mutations on the genes encoding histone proteins are also linked with cancers. 30% of paediatric glioblastomas have mutations at key post translational modification sites in histone genes [93]. Recently, mutations in metabolic enzymes have also been observed to have a role in histone methylation status alteration. The mutated metabolic enzymes produce altered metabolites (popularly known as oncometabolites) which jeopardize the function of methylase enzymes. For instance, inhibition of histone demethylation Jumonji C enzymes by the oncometabolite d-2-hydroxyglutarate [94–97].

4.1.5 Histone phosphorylation

Phosphorylation of histones takes place on serine, threonine, tyrosine and histidine residues, predominantly in the N-terminal tails of all nucleosomal histones by histone kinase enzymes which transfer a phosphate group from ATP to the hydroxyl group of the target amino-acid side chain. Phosphate group contains significant negative charge and therefore phosphorylation is generally associated with transcriptional upregulation. Various proteins have been identified which contain phosphor-binding domains [98, 99]. Histone phosphorylation changes dynamically with the transcriptional profile of the cell [100]. For example, H3Ser10 phosphorylation correlates with gene activation in mammalian cells and heat shock response induced transcription in *Drosophila* [101]. However, the same phosphorylation is associated with chromosome condensation and segregation during mitosis and meiosis [102]. Histone phosphorylations also play a pivotal role in response to DNA damage e.g., phosphorylation of H2A(X) on serine 139 in mammalian cells (referred to as γ H2AX) and S129 of H2A in yeast [103].

4.1.6 Histone phosphorylation and cancer

Regulation of the level of histone H3 phosphorylation by an interplay of the activities of kinases and phosphatases serves as a means of promoting chromosomal condensation and segregation in mitosis [104]. Phosphorylation of H3S10 has also been linked to the expression of proto-oncogenes like *c-fos* [105–107]. It has been detected with the aid of ChIP assay that phosphoacetylation of H3 tails exist at the promoters of several MAP- kinase activated genes as well as the promoters of *c-fos* and *c-jun* [108]. H2A(X) phosphorylation is involved breast cancer [109] and colon cancer [110]. Histidine phosphorylation on histone H4 has been shown to be involved in liver regeneration and cancer [111]. Phosphoacetylation of histones, involving phosphorylation of histone H3 on residue serine 10 and acetylation of histone H4 on lysine 12 has been shown to have a role prognosis of oral squamous cell carcinoma [112].

4.1.7 Histone ubiquitination

It is a process in which ubiquitin molecules are added to lysine residues of histones. Monoubiquitination is the major form of ubiquitination in histones. However, histones H2A and H2B can also be modified by polyubiquitination. The first ubiquitinated histone to be identified was H2A [113]. H2A and H2B also hold the distinction of being the most abundantly ubiquitinated proteins in the nucleus [113, 114]. In addition, H3, H4 as well as H1 have been reported to be modified by ubiquitin but the biological function of these ubiquitinations has not been well characterised [115, 116]. Histone ubiquitinations perform a number of important nucleosomal functions. Chromatin immunoprecipitation (ChIP) experiments have revealed enrichment of monoubiquitinated H2A (H2Aub) in the satellite regions of genome and of H2Bub in transcriptionally active genes [117, 118].

4.1.8 Histone ubiquitination and cancer

Several recent studies have linked ubiquitination, especially H2Bub with inflammation and cancer [119–121]. Histone H2Bub1 predominantly resides downstream to transcription start sites (TSS), a position which allows association with highly transcribed genes and therefore makes this protein a likely target in cancer [117]. RNF20/RNF40 has been shown to negatively regulate cancer- related inflammation in mice and humans through increased recruitment of repressive NF-κB subunit p50 to various gene targets to downregulate their transcription [121]. RNF40 is also known to modulate NF-κB activity in colorectal cancer in mice [122] while as RNF20 and H2B ubiquitylation have also been shown to be involved in breast cancer [123]. Loss of H2B monoubiquitination has also been shown to activate immune pathways by alteration of chromatin accessibility in ovarian cancer [124–126].

5. Conclusion

Epigenome in a typical eukaryotic cell is packaged as an entity containing nucleo-proteins-DNA and histones. This epigenome is compartmentalized into euchromatin and heterochromatin and contain various marks which are transmitted from one cell generation to another [127]. Covalent DNA and histone modifications are the carriers of epigenetic inheritance which are required for the maintenance of a stable epigenome [128]. Any disturbance in the propagation and maintenance of a stable epigenome is associated with diseases like transformation and cancer. The process of cellular transformation is associated with changes in the epigenetic landscape of DNA methylation and histone post-translational modifications. In recent past, genome wide studies have identified various genes related to diseases like cancer and neuro-degeneration [4]. Many of these genes have been observed to code for key epigenetic enzymes like HDACs, which raises the possibility of their involvement in far reaching pathological problems. In recent years, non-coding RNA has also been increasingly investigated in relation to carcinogenesis and various types of non-coding RNAs have been associated with different forms of cancer [129, 130].

A stable epigenome also requires proper chromatin conformation. It has been observed that upon transformation, the 3D organization and nuclear topology also undergoes certain changes. These topological changes can be both cause and consequence of alterations in histone and DNA modifications. Topological changes in chromatin structure are associated with increased expression of repetitive DNA elements, which leads to hyper-recombination and gross genomic instability which can further lead a cell on the path of transformation.

Studies performed on chromatin structure and covalent modifications have paved way for better understanding as well as therapeutic intervention of various forms of cancer. Epigenetic approach of therapeutic intervention in cancer is definitely a better approach for cancer treatment since it aims at reversal of inheritable changes without changing the DNA or without affecting normal physiological processes. Also, tumor forms have recently been discovered with anatomical restrictions which contain mutations in histone variant genes. For example, H3.3, a variant of histone H3, contains a point mutation at residue 34 in which glycine changes to valine or arginine (H3.3G34V or H3.3G34R). These tumors are found almost exclusively in the cerebral hemispheres [131, 132]. Tumors with point mutations in histone variant H3.1 (H3.1K27M) are restricted to pons of brainstem while as H3.3K27M tumors are found along the midline of the brain [133]. This "anatomical restriction" in tumor types and the corresponding mutations in histone variants are indicative of an excit-ing new dimension of the role of epigenetics in tumor biology [134, 135]. This also provides cues about the role of epigenetics in defining tumor micro-environment. Alternatively, many more tumor types can be screened for mutations in genes coding

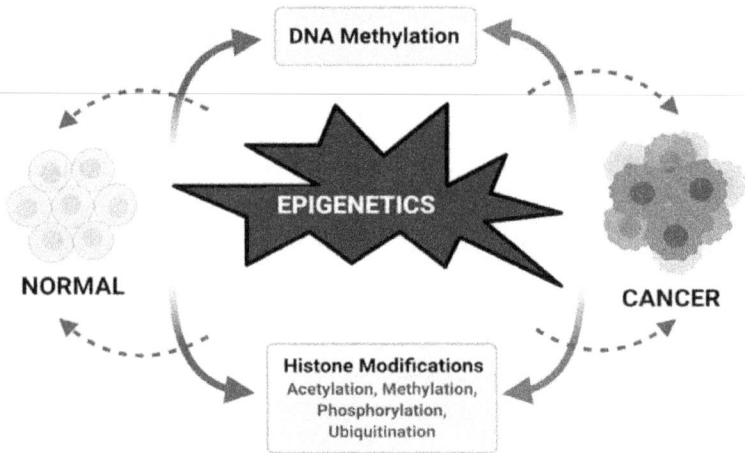

Figure 4.
Schematic depicting two major pillars of epigenetic mechanisms that is, DNA methylation and histone modifications, their importance in maintaining normal cellular morphology and function and their mis-regulation leading to cancer.

for epigenetic factors to have better insights into the role of epigenetics in tumor progression. These findings also encourage the possibility of exploration of epigenetic therapy in resetting the balance in tumor micro-environment for therapeutic targeting. However, the field of epigenetic studies and epigenetic cancer therapy is still in its infancy and intense investigations are required for further exploration of the possibility of epigenetic targeting and treatment of cancer (**Figure 4**).

Author details

Zeenat Farooq[1*], Ambreen Shah[2], Mohammad Tauseef[3], Riyaz Ahmad Rather[4] and Mumtaz Anwar[1*]

1 Department of Pharmacology and Regenerative Medicine, College of Medicine, University of Illinois at Chicago, Chicago, USA

2 Department of Biotechnology, University of Kashmir, India

3 Department of Pharmaceutical Sciences, College of Pharmacy, Chicago State University, Chicago, USA

4 Department of Biotechnology, Wachemo University, Ethiopia

*Address all correspondence to: zeenatfa@uic.edu and mumtazan@uic.edu

IntechOpen

References

[1] Farooq Z, et al. The many faces of histone H3K79 methylation. Mutat Res Rev Mutat Res. 2016 Apr-Jun;768:46-52.

[2] Altaf M, et al. Histone modifications in response to DNA damage. Mutat Res. 2007 May 1;618(1-2):81-90.

[3] Banday S, et al. Role of Inner Nuclear Membrane Protein Complex Lem2-Nur1 in Heterochromatic Gene Silencing. J Biol Chem. 2016 Sep 16;291(38):20021-9.

[4] Ganai SA, et al. Modulating epigenetic HAT activity for reinstating acetylation homeostasis: A promising therapeutic strategy for neurological disorders. Pharmacol Ther. 2016 Oct;166:106-22.

[5] Banday S, et al. Therapeutic strategies against hDOT1L as a potential drug target in MLL-rearranged leukemias. Clin Epigenetics. 2020 May 25;12(1):73.

[6] Ganai SA, et al. In silico approaches for investigating the binding propensity of apigenin and luteolin against class I HDAC isoforms. Future Med Chem. 2018 Aug 1;10(16):1925-1945.

[7] Robertson KD. DNA methylation, methyltransferases, and cancer. Oncogene. 2001 May 28;20(24):3139-55.

[8] Bestor TH. The DNA methyltransferases of mammals. Hum Mol Genet. 2000 Oct;9(16):2395-402.

[9] Bird A. DNA methylation patterns and epigenetic memory. Genes Dev. 2002 Jan 1;16(1):6-21.

[10] Okano M, et al. DNA methyltransferases Dnmt3a and Dnmt3b are essential for de novo methylation and mammalian development. Cell. 1999 Oct 29;99(3):247-57.

[11] Friedman RC, et al. Most mammalian mRNAs are conserved targets of microRNAs. Genome Res. 2009 Jan;19(1):92-105.

[12] Deplus R, et al. Dnmt3L is a transcriptional repressor that recruits histone deacetylase. Nucleic Acids Res. 2002 Sep 1;30(17):3831-8.

[13] Robertson KD. DNA methylation and human disease. Nat Rev Genet. 2005 Aug;6(8):597-610.

[14] Kim JK, et al. Epigenetic mechanisms in mammals. Cell Mol Life Sci. 2009 Feb;66(4):596-612.

[15] Jones PA, and Laird PW. Cancer epigenetics comes of age. Nat Genet. 1999 Feb;21(2):163-7.

[16] Jones PA, and Baylin SB. The fundamental role of epigenetic events in cancer. Nat Rev Genet. 2002 Jun;3(6):415-28.

[17] Lander ES, et al. Initial sequencing and analysis of the human genome. Nature. 2001 Feb 15;409(6822):860-921.

[18] Bogdanovic O, and Veenstra GJ. DNA methylation and methyl-CpG binding proteins: developmental requirements and function. Chromosoma. 2009 Oct;118(5): 549-565.

[19] Dehan P, et al. DNA methylation and cancer diagnosis: new methods and applications. Expert Rev Mol Diagn. 2009 Oct;9(7):651-7.

[20] Lai AY, and Wade PA. Cancer biology and NuRD: a multifaceted chromatin remodelling complex. Nat Rev Cancer. 2011 Jul 7;11(8):588-96.

[21] Du Q, et al. Methyl-CpG-binding domain proteins: readers of the epigenome. Epigenomics. 2015;7(6): 1051-73.

[22] Shiah SG, et al. The involvement of promoter methylation and DNA methyltransferase-1 in the regulation of EpCAM expression in oral squamous cell carcinoma. Oral Oncol. 2009 Jan;45(1):e1-8.

[23] Zhu YM, et al. Expression of human DNA methyltransferase 1 in colorectal cancer tissues and their corresponding distant normal tissues. Int J Colorectal Dis. 2007 Jun;22(6):661-6.

[24] Rhee I, et al. CpG methylation is maintained in human cancer cells lacking DNMT1. Nature. 2000 Apr 27;404(6781):1003-7.

[25] Ting AH, et al. Mammalian DNA methyltransferase 1: inspiration for new directions. Cell Cycle. 2004 Aug;3(8): 1024-6.

[26] Kanai Y, et al. Aberrant DNA methylation precedes loss of heterozygosity on chromosome 16 in chronic hepatitis and liver cirrhosis. Cancer Lett. 2000 Jan 1;148(1):73-80.

[27] Valinluck V, and Sowers LC. Endogenous cytosine damage products alter the site selectivity of human DNA maintenance methyltransferase DNMT1. Cancer Res. 2007 Feb 1;67(3):946-50.

[28] Feinberg AP, Tycko B. The history of cancer epigenetics. Nat Rev Cancer. 2004 Feb;4(2):143-53.

[29] Herman JG, and Baylin SB. Gene silencing in cancer in association with promoter hypermethylation. N Engl J Med. 2003 Nov 20;349(21):2042-54.

[30] Daura-Oller E, et al. Specific gene hypomethylation and cancer: New insights into coding region feature trends. Bioinformation. 2009; 3(8): 340-343.

[31] Arai E, et al. Regional DNA hypermethylation and DNA methyltransferase (DNMT) 1 protein overexpression in both renal tumors and corresponding nontumorous renal tissues. nt J Cancer. 2006 Jul 15;119(2):288-96.

[32] Ogi K, et al. Aberrant methylation of multiple genes and clinicopathological features in oral squamous cell carcinoma. Clin Cancer Res. 2002 Oct;8(10):3164-71.

[33] Kulkarni V, and Saranath D. Concurrent hypermethylation of multiple regulatory genes in chewing tobacco associated oral squamous cell carcinomas and adjacent normal tissues. Oral Oncol. 2004 Feb;40(2):145-53.

[34] Shaw RJ, et al. Quantitative methylation analysis of resection margins and lymph nodes in oral squamous cell carcinoma. Br J Oral Maxillofac Surg. 2007 Dec;45(8):617-22.

[35] Rosas SL, et al. Promoter hypermethylation patterns of p16, O6-methylguanine-DNA-methyltransferase, and death-associated protein kinase in tumors and saliva of head and neck cancer patients. Cancer Res. 2001 Feb 1;61(3):939-42.

[36] Lopez M, et al. Gene promoter hypermethylation in oral rinses of leukoplakia patients—a diagnostic and/ or prognostic tool? Eur J Cancer. 2003 Nov;39(16):2306-9.

[37] Oh BK, et al. DNA methyltransferase expression and DNA methylation in human hepatocellular carcinoma and their clinicopathological correlation. Int J Mol Med. 2007 Jul;20(1):65-73.

[38] Sanchez-Cespedes M, et al. Gene promoter hypermethylation in tumors and serum of head and neck cancer patients. Cancer Res. 2000 Feb 15;60(4):892-5.

[39] Maruya S, et al. Differential methylation status of tumor-associated genes in head and neck squamous carcinoma: incidence and potential implications. Clin Cancer Res. 2004 Jun 1;10(11):3825-30.

[40] Eads CA, et al. CpG island hypermethylation in human colorectal tumors is not associated with DNA methyltransferase overexpression. Cancer Res. 1999 May 15;59(10):2302-6. Erratum in: Cancer Res 1999 Nov 15;59(22):5860.

[41] Arnold CN, et al. Evaluation of microsatellite instability, hMLH1 expression and hMLH1 promoter hypermethylation in defining the MSI phenotype of colorectal cancer. Cancer Biol Ther. 2004 Jan;3(1):73-8.

[42] Malhotra P, et al. Promoter methylation and immunohistochemical expression of hMLH1 and hMSH2 in sporadic colorectal cancer: a study from India. Tumour Biol. 2014 Apr;35(4): 3679-87.

[43] Kim H, et al. Elevated mRNA levels of DNA methyltransferase-1 as an independent prognostic factor in primary non-small cell lung cancer. Cancer. 2006 Sep 1;107(5):1042-9.

[44] Dansranjavin T, Möbius C, Tannapfel A, Bartels M, Wittekind C, Hauss J, Witzigmann H. E-cadherin and DAP kinase in pancreatic adenocarcinoma and corresponding lymph node metastases. Oncol Rep. 2006 May;15(5):1125-31.

[45] Shieh YS, et al. DNA methyltransferase 1 expression and promoter methylation of E-cadherin in mucoepidermoid carcinoma. Cancer. 2005 Sep 1;104(5):1013-21.

[46] Oshimo Y, et al. Promoter methylation of cyclin D2 gene in gastric carcinoma. Int J Oncol. 2003 Dec;23(6):1663-70.

[47] De Smet C, et al. Promoter-dependent mechanism leading to selective hypomethylation within the 5' region of gene MAGE-A1 in tumor cells. Mol Cell Biol. 2004 Jun;24(11):4781-90.

[48] Fendri A, et al. Inactivation of RASSF1A, RARbeta2 and DAP-kinase by promoter methylation correlates with lymph node metastasis in nasopharyngeal carcinoma. Cancer Biol Ther. 2009 Mar;8(5):444-51.

[49] Friedrich MG, et al. Detection of methylated apoptosis-associated genes in urine sediments of bladder cancer patients. Clin Cancer Res. 2004 Nov 15;10(22):7457-65.

[50] Abbosh PH, et al. Hypermethylation of tumor-suppressor gene CpG islands in small-cell carcinoma of the urinary bladder. Mod Pathol. 2008 Mar;21(3):355-62.

[51] Vallböhmer D, et al. DNA methyltransferases messenger RNA expression and aberrant methylation of CpG islands in non-small-cell lung cancer: association and prognostic value. Clin Lung Cancer. 2006 Jul;8(1):39-44.

[52] Hong, SM, et al. Loss of E-cadherin expression and outcome among patients with resectable pancreatic adenocarcinomas. Mod Pathol. 2011 Sep;24(9):1237-47.

[53] Christoph F, et al. mRNA expression profiles of methylated APAF-1 and DAPK-1 tumor suppressor genes uncover

clear cell renal cell carcinomas with aggressive phenotype. J Urol. 2007 Dec;178(6):2655-9.

[54] Mumtaz Anwar, et al. Mutations & expression of APC & β-Catenin in sporadic colorectal tumors: A mutational "hotspot" for tumorigenesis. Journal of Gastroenterology and Hepatology. 2013; 28: 665-666.

[55] Bonenfant D, et al. Characterization of histone H2A and H2B variants and their post-translational modifications by mass spectrometry. Mol Cell Proteomics. 2006 Mar;5(3):541-52.

[56] Wyrick JJ, et al. The role of histone H2A and H2B post-translational modifications in transcription: a genomic perspective. Biochim Biophys Acta. 2009 Jan;1789(1):37-44.

[57] Ning Song, et al. Immunohistochemical Analysis of Histone H3 Modifications in Germ Cells during Mouse Spermatogenesis. Acta Histochem Cytochem. 2011 Aug 27;44(4):183-90.

[58] Berger SL. Histone modifications in transcriptional regulation. Curr Opin Genet Dev. 2002 Apr;12(2):142-8.

[59] Kouzarides T. Histone methylation in transcriptional control. Curr Opin Genet Dev. 2002 Apr;12(2):198-209. doi: 10.1016/s0959-437x(02)00287-3. Erratum in: Curr Opin Genet Dev 2002 Jun;12(3):371.

[60] Dhalluin C, et al. Structure and ligand of a histone acetyltransferase bromodomain. Nature. 1999 Jun 3;399(6735):491-6.

[61] Jacobs SA and Khorasanizadeh S. Structure of HP1 chromodomain bound to a lysine 9-methylated histone H3 tail. Science. 2002 Mar 15;295(5562):2080-3.

[62] Strahl BD, Allis CD. The language of covalent histone modifications. Nature. 2000 Jan 6;403(6765):41-5.

[63] Fraga MF, et al. Loss of acetylation at Lys16 and trimethylation at Lys20 of histone H4 is a common hallmark of human cancer. Nat Genet. 2005 Apr;37(4):391-400.

[64] Gui CY, et al. Histone deacetylase (HDAC) inhibitor activation of p21WAF1 involves changes in promoter-associated proteins, including HDAC1. Proc Natl Acad Sci U S A. 2004 Feb 3;101(5):1241-6.

[65] Yasui W, et al. Histone acetylation and gastrointestinal carcinogenesis. Ann N Y Acad Sci. 2003 Mar;983:220-31.

[66] Lin RJ, Sternsdorf T, Tini M, Evans RM. Transcriptional regulation in acute promyelocytic leukemia. Oncogene. 2001 Oct 29;20(49):7204-15.

[67] Christofori G, Semb H. The role of the cell-adhesion molecule E-cadherin as a tumour-suppressor gene. Trends Biochem Sci. 1999 Feb;24(2):73-6.

[68] Anwar M, et al. Frequent activation of the β-catenin gene in sporadic colorectal carcinomas: A mutational & expression analysis. Mol Carcinog. 2016 Nov;55(11):1627-1638.

[69] Choi JH, et al. Expression profile of histone deacetylase 1 in gastric cancer tissues. Jpn J Cancer Res. 2001 Dec;92(12):1300-4.

[70] Halkidou K, et al. Upregulation and nuclear recruitment of HDAC1 in hormone refractory prostate cancer. Prostate. 2004 May 1;59(2):177-89.

[71] Wilson AJ, et al. Histone deacetylase 3 (HDAC3) and other class I HDACs regulate colon cell maturation and p21

expression and are deregulated in human colon cancer. J Biol Chem. 2006 May 12;281(19):13548-13558.

[72] Zhang Z, et al. Quantitation of HDAC1 mRNA expression in invasive carcinoma of the breast. Breast Cancer Res Treat. 2005 Nov;94(1):11-6.

[73] Huang BH, et al. Inhibition of histone deacetylase 2 increases apoptosis and p21Cip1/WAF1 expression, independent of histone deacetylase 1. Cell Death Differ. 2005 Apr;12(4):395-404.

[74] Song J, et al. Increased expression of histone deacetylase 2 is found in human gastric cancer. APMIS. 2005 Apr;113(4):264-8.

[75] Zhu P, et al. Induction of HDAc2 expression upon loss of APC in colorectal tumorogenesis. Cancer Cell. 2004 May;5(5):455-63.

[76] Zhang Z, et al. HDAC6 expression is correlated with better survival in breast cancer. Clin Cancer Res. 2004 Oct 15;10(20):6962-8.

[77] Ropero S, et al. A truncating mutation of HDAC2 in human cancers confers resistance to histone deacetylase inhibition. Nat Genet. 2006 May;38(5):566-9.

[78] Frolov MV, Dyson NJ. Molecular mechanisms of E2F-dependent activation and pRB-mediated repression. J Cell Sci. 2004 May 1;117(Pt 11):2173-81.

[79] Vaziri H, et al. hSIR2(SIRT1) functions as an NAD-dependent p53 deacetylase. Cell. 2001 Oct 19;107(2):149-59.

[80] Fu M, et al. Hormonal control of androgen receptor function through SIRT1. Mol Cell Biol. 2006 Nov;26(21): 8122-35.

[81] Bouras T, et al. SIRT1 deacetylation and repression of p300 involves lysine residues 1020/1024 within the cell cycle regulatory domain 1. J Biol Chem. 2005 Mar 18;280(11):10264-76.

[82] Wang C, et al. Interactions between E2F1 and SirT1 regulate apoptotic response to DNA damage. Nat Cell Biol. 2006 Sep;8(9):1025-31.

[83] Cohen HY, et al. Calorie restriction promotes mammalian cell survival by inducing the SIRT1 deacetylase. Science. 2004 Jul 16;305(5682):390-2.

[84] Yeung F, et al. Modulation of NF-kappaB dependent transcription and cell survival by the SIRT1 deacetylase. EMBO J. 2004 Jun 16;23(12):2369-80.

[85] Yu Wang and Jia Songtao. Degrees make all the difference. Epigenetics. 2009 Jul 1; 4(5): 273-276.

[86] Scott F. Gilbert-Developmental Biology, 2010, Sinauer Associates, Inc., Sunderland, MA Ninth Edition, September 2011 Russian Journal of Developmental Biology 42(5)

[87] Takagi N, Sasaki M. Preferential inactivation of the paternally derived X chromosome in the extraembryonic membranes of the mouse. Nature. 1975 Aug 21;256(5519):640-2.

[88] Elgin SC, Grewal SI. Heterochromatin: silence is golden. Curr Biol. 2003 Dec 2;13(23):R895-8.

[89] Jin Y, et al. Simultaneous Mutation of Methylated Lysine Residues in Histone H3 Causes Enhanced Gene Silencing, Cell Cycle Defects, and Cell Lethality in Saccharomyces Cerevisiae. Mol Cell Biol. 2007 Oct;27(19):6832-41.

[90] Esteller M. Epigenetics provides a new generation of oncogenes and

tumour-suppressor genes. Br J Cancer. 2007;96 Suppl:R26-30.

[91] Dawson MA, Kouzarides T. Cancer epigenetics: from mechanism to therapy. Cell. 2012 Jul 6;150(1):12-27.

[92] You JS, Jones PA. Cancer genetics and epigenetics: two sides of the same coin? Cancer Cell. 2012 Jul 10;22(1):9-20.

[93] Mohammad F & Helin K. Oncohistones: drivers of pediatric cancers. Genes Dev. 31(23-24):2313-2324.

[94] Garrett-Bakelman FE, and Melnick AM. Mutant IDH: a targetable driver of leukemic phenotypes linking metabolism, epigenetics and transcriptional regulation. Epigenomics. 2016 Jul;8(7):945-57.

[95] Elkashef SM, et al. IDH mutation, competitive inhibition of FTO, and RNA methylation. Cancer Cell. 2017 May 8;31(5):619-620.

[96] Flavahan WA, et al. Insulator dysfunction and oncogene activation in IDH mutant gliomas. Nature. 2016 Jan 7;529(7584):110-4.

[97] Xu W et al. Oncometabolite 2-hydroxyglutarate is a competitive inhibitor of alpha-ketoglutarate-dependent dioxygenases. Cancer Cell. 2011 Jan 18;19(1):17-30.

[98] Taverna SD, et al. How chromatin-binding modules interpret histone modifications. Nat Struct Mol Biol. 2007 Nov;14(11):1025-1040.

[99] Yun M, et al. Readers of histone modifications. Cell Res. 2011 21:564-578.

[100] Nowak SJ, and Corces VG. Phosphorylation of histone H3 correlates with transcriptionally active loci. Genes Dev. 2000 Dec 1;14(23):3003-13.

[101] Sassone-Corsi P, et al. Requirement of Rsk-2 for epidermal growth factor-activated phosphorylation of histone H3. Science. 1999 Aug 6;285(5429):886-91.

[102] Gurley LR, et al. Histone phosphorylation and chromatin structure during mitosis in Chinese hamster cells. Eur J Biochem. 1978 Mar;84(1):1-15.

[103] Nagelkerke A, and Span PN. Staining Against Phospho-H2AX (gamma-H2AX) as a Marker for DNA Damage and Genomic Instability in Cancer Tissues and Cells. Adv Exp Med Biol. 2016;899:1-10.

[104] Hans F, Dimitrov S. Histone H3 phosphorylation and cell division. Oncogene. 2001 May 28;20(24):3021-7.

[105] Fernandez-Capetillo O, et al. H2AX: the histone guardian of the genome. DNA Repair (Amst). 2004 Aug-Sep;3(8-9):959-67.

[106] Cheung WL, et al. Phosphorylation of histone H4 serine 1 during DNA damage requires casein kinase II in S. Cerevisiae. Curr Biol. 2005 Apr 12;15(7):656-60.

[107] Utley RT, et al Regulation of NuA4 histone acetyltransferase activity in transcription and DNA repair by phosphorylation of histone H4. Mol Cell Biol. 2005 Sep;25(18):8179-90.

[108] Ajiro K. Histone H2B phosphorylation in mammalian apoptotic cells. An association with DNA fragmentation. J Biol Chem. 2000 Jan 7;275(1):439-43.

[109] Liu Y, et al. JMJD6 regulates histone H2A.X phosphorylation and promotes autophagy in triple-negative breast cancer cells via a novel tyrosine kinase activity. Oncogene. 2019 Feb;38(7): 980-997.

[110] Liu Z, et al. EZH2 regulates H2B phosphorylation and elevates colon cancer cell autophagy. J Cell Physiol. 2020 Feb;235(2):1494-1503.

[111] Besant PG, and Attwood PV. Histone H4 histidine phosphorylation: kinases, phosphatases, liver regeneration and cancer. Biochem Soc Trans. 2012 Feb;40(1):290-3.

[112] Campos-Fernández E, et al. Prognostic value of histone H3 serine 10 phosphorylation and histone H4 lysine 12 acetylation in oral squamous cell carcinoma. Histopathology. 2019 Jan;74(2):227-238.

[113] Goldknopf IL, et al. Isolation and characterization of protein A24, a 'histone-like' non-histone chromosomal protein. J Biol Chem. 1975 Sep 25;250(18):7182-7.

[114] West MH, and Bonner WM. Histone 2B can be modified by the attachment of ubiquitin. Nucleic Acids Res. 1980 Oct 24;8(20):4671-80.

[115] Pham AD, and Sauer F. Ubiquitin-activating/conjugating activity of TAFII250, a mediator of activation of gene expression in Drosophila. Science. 2000 Sep 29;289(5488):2357-60.

[116] Jones JM, et al. The RAG1V(D)J recombinase/ubiquitin ligase promotes ubiquitylation of acetylated, phosphorylated histone3.3. Immunol Lett. 2011 May;136(2):156-62.

[117] Minsky N, et al. Monoubiquitinated H2B is associated with the transcribed region of highly expressed genes in human cells. Nat Cell Biol. 2008 Apr;10(4):483-8.

[118] Zhu Q, et al. BRCA1 tumour suppression occurs via heterochromatin-mediated silencing. Nature. 2011 Sep 7;477(7363):179-84.

[119] Sethi G, et al. Role of RNF20 in cancer development and progression - a comprehensive review. Biosci Rep. 2018 Jul 12;38(4):BSR20171287.

[120] Cole AJ, et al. Histone H2B monoubiquitination: roles to play in human malignancy. Endocr Relat Cancer. 2015 Feb;22(1):T19-33.

[121] Tarcic O, et al. RNF20 Links Histone H2B Ubiquitylation with Inflammation and Inflammation-Associated Cancer. Cell Rep. 2016 Feb 16;14(6):1462-1476.

[122] Kosinsky RL, et al. Loss of RNF40 Decreases NF-κB Activity in Colorectal Cancer Cells and Reduces Colitis Burden in Mice. J Crohns Colitis. 2019 Mar 26;13(3):362-373.

[123] Tarcic O, et al. RNF20 and histone H2B ubiquitylation exert opposing effects in Basal-Like versus luminal breast cancer. Cell Death Differ. 2017 Apr;24(4):694-704.

[124] Hooda J, et al. Early Loss of Histone H2B Monoubiquitylation Alters Chromatin Accessibility and Activates Key Immune Pathways That Facilitate Progression of Ovarian Cancer. Cancer Res. 2019 Feb 15;79(4):760-772.

[125] Dickson KA, et al. The RING finger domain E3 ubiquitin ligases BRCA1 and the RNF20/RNF40 complex in global loss of the chromatin mark histone H2B monoubiquitination (H2Bub1) in cell line models and primary high-grade serous ovarian cancer. Hum Mol Genet. 2016 Dec 15;25(24):5460-5471.

[126] Marsh DJ, et al. Histone Monoubiquitination in Chromatin Remodelling: Focus on the Histone H2B Interactome and Cancer. Cancers (Basel). 2020 Nov 20;12(11):3462.

[127] Farooq Z, et al. Vigilin protein Vgl1 is required for heterochromatin-mediated gene silencing in

Schizosaccharomyces pombe. J
Biol Chem. 2019 Nov 29;294(48):
18029-18040.

[128] Tan L, et al. Naked mole rat cells
have a stable epigenome that resists iPSC
reprogramming. Stem Cell Reports. 2017
Nov 14;9(5):1721-1734.

[129] Michalak EM, et al. The roles of
DNA, RNA and histone methylation in
ageing and cancer. Nat Rev Mol Cell Biol.
2019 Oct;20(10):573-589.

[130] Anastasiadou E, et al. Non-coding
RNA networks in cancer. Nat Rev Cancer.
2018 Jan;18(1):5-18.

[131] Schwartzentruber J, et al. Driver
mutations in histone H3.3 and chromatin
remodelling genes in paediatric
glioblastoma. Nature. 2012 Jan
29;482(7384):226-31.

[132] Sturm D, et al. Hotspot mutations in
H3F3A and IDH1 define distinct
epigenetic and biological subgroups of
glioblastoma. Cancer Cell. 2012 Oct
16;22(4):425-37.

[133] Mackay A, et al. Integrated
molecular meta-analysis of 1,000
pediatric high-grade and diffuse intrinsic
pontine glioma. Cancer Cell. 2017 Oct
9;32(4):520-537.e5.

[134] Mohammad HP, et al. Targeting
epigenetic modifications in cancer
therapy: erasing the roadmap to cancer.
Nat Med. 2019 Mar;25(3):403-418.

[135] Bennett RL, and Licht JD. Targeting
Epigenetics in Cancer. Annu Rev
Pharmacol Toxicol. 2018 Jan 6;58:187-207.

Chapter 3

Diabetes and Epigenetics

Rasha A. Alhazzaa, Thomas Heinbockel and Antonei B. Csoka

Abstract

As we attempt to understand and treat diseases, the field of epigenetics is receiving increased attention. For example, epigenetic changes may contribute to the etiology of diabetes. Herein, we review the histology of the pancreas, sugar metabolism and insulin signaling, the different types of diabetes, and the potential role of epigenetic changes, such as DNA methylation, in diabetes etiology. These epigenetic changes occur at differentially-methylated sites or regions and have been previously linked to metabolic diseases such as obesity. In particular, changes in DNA methylation in cells of the pancreatic islets of Langerhans may be linked to type 2 diabetes (T2D), which in turn is related to peripheral insulin resistance that may increase the severity of the disease. The hypothesis is that changes in the epigenome may provide an underlying molecular mechanism for the cause and deleterious metabolic health outcomes associated with severe obesity or T2D. Conversely, reversing such epigenetic changes may help improve metabolic health after therapeutic interventions.

Keywords: glucose, pancreas, beta cells, epigenetics, DNA methylation, diabetes mellitus, hypomethylation, hypermethylation, histone modification

1. Introduction

Diabetes mellitus (DM) is a chronic metabolic disease in which either the pancreas produces very little to no insulin, termed type 1 diabetes (T1D), or insufficient insulin in the context of systemic insulin resistance, termed type 2 diabetes (T2D) [1]. Both of these conditions result in high levels of glucose in the bloodstream.

DM is associated with significantly elevated diabetic nephropathy, neuropathy, and retinopathy, which are microvascular complications, and cardiovascular conditions such as hypertension, atherosclerosis, and stroke, which are considered macrovascular diseases. DM is associated with genetic as well as environmental factors, with the cost of treatment and debilitating complications increasing dramatically due to an epidemic of DM worldwide.

However, the above statement is something of an oversimplification, because besides T1D and T2D, there are even more variants, and we will now look at all of these in more detail.

2. Type 1 diabetes

T1D represents only around 10% of the DM cases worldwide but is increasingly occurring earlier in life. It results from autoimmune destruction of the beta cells (β-cells) of the endocrine pancreas. As with cancer, obesity, and autoimmune diseases, T1D results from the interaction of genetic and non-genetic factors [2]. In autoimmune diseases, due to an immunological malfunction and lack of tolerance of self-antigens, the immune system destroys the body's own tissues. More than 80 different diseases are considered autoimmune and affect approximately 100 million people worldwide [3]. T1D is one such example that results in the slow degeneration and destruction of the pancreas. However, approximately 10% of the affected patients are classified as subtype 1B, and the pathogenesis in these cases is considered idiopathic since there is no evidence of autoimmunity [4]. T1D can be correlated to ethnicity, gender, genetics, and environmental influences. For instance, it occurs more in children and those under the age of 20 and affects both male and female children equally. However, studies have found that males are disproportionally affected in areas with a high prevalence of T1D, whereas females are disproportionally affected in areas with a low prevalence of the disease. It is highest in non-Hispanic White people and lowest in Navajo groups. Moreover, T1D is common in families with a history of the disease. Epidemiological studies have found an association between T1D and environmental factors, and dietary and nutritional habits [5].

Essential mediators leading to β-cell destruction in T1D include the following pro-inflammatory cytokines: interleukin-1β (IL-1β); interferon-γ (IFN-γ); and tumor necrosis factor-α (TNF-α). These cytokines induce the overexpression of iNOS in β-cells, leading to an overproduction of NO that causes cytotoxicity. This suggests an important role for NO in the pathogenesis of DM [6].

3. Type 2 diabetes

In contrast to T1D, T2D is a defect in insulin secretion, insulin action, or both that leads to the development of a multifactorial and heterogeneous group of disorders. Changes in diet and physical activity levels have led to an increased worldwide prevalence of T2D over the past several decades. There is also strong evidence supporting a genetic component of T2D susceptibility, and several genes underlying monogenic forms of DM have already been identified. However, T2D likely results from the contribution of many genes interacting with different environmental factors to produce wide variations in the clinical course [7].

Regarding this process, there is a decrease in β-cell mass in T2D with the primary implicated mechanism being the apoptosis of the cells. This type of dynamic cell death is increased in all diabetic individuals; β-cell mass depends on many factors, including cell size, cell renewal rate from proliferation of pre-existing cells or neogenesis (differentiation from other precursor cells), and speed of apoptosis. Also, β-cell failure during the progression to T2D can be caused by either chronic exposure of the β-cell to glucose, which is called "glucotoxicity," or exposure to fatty acids, which is known as "lipotoxicity" [8].

4. Latent autoimmune diabetes in adults and maturity-onset diabetes of the young

Besides T1D and T2D, some forms of the disease do not fit neatly into those groups, namely latent autoimmune diabetes in adults (LADA) and maturity-onset diabetes of the young (MODY). LADA shares some type 1 and type 2 symptoms and treatments, is diagnosed during adulthood, and sets in gradually, like T2D [9]. On the other hand, MODY is caused by genetic changes that affect how well the body makes insulin [10].

5. Type 3 diabetes

There may be a connection between T2D and Alzheimer's disease (AD). Indeed, it has been proposed that AD is actually a form of DM termed type 3 (T3D). Globally, the epidemic of T2D and the possibility of it contributing to the risk of AD have become a paramount health concern.

The hypothesis is that T3D corresponds to chronic insulin resistance plus an insulin-deficient state that is mostly confined to the brain [11, 12]. A deficit in glucose utilization is observed, which ultimately leads to cognitive dysfunction. Over time T3D steadily destroys cerebral functions due to insulin imbalance. Thus, the central nervous system develops insulin resistance, which leads to AD [13].

6. Gestational diabetes

Finally, in terms of DM classifications, the American Diabetes Association defines gestational DM (GDM) as DM seen during pregnancy. GDM occurs in approximately 5% of pregnancies, but rates can increase due to obesity. Pregnancies with a diagnosis of GDM present a risk to both mother and child. Women who have a record of GDM will typically develop T2D after pregnancy. Their children have a higher incidence of becoming obese and developing T2D early in life [14].

So, as we can see, ultimately, DM and its resultant health conditions are numerous and include many degenerative diseases such as the above-described six types of DM.

But possibly, this myriad of insidious conditions and diseases could be preventable if we focus research on the pancreatic β-cells (vital to the regulation of glucose levels in the bloodstream). But first, what exactly are β-cells? At this point, we will look at the histological composition of the pancreas in more detail.

7. The cells of the pancreas

Histologically, the adult pancreas consists of endocrine and exocrine cells, but these cells can change their state of differentiation in response to various stimuli (e.g., injury or stress). The *exocrine* portion of the pancreas produces and releases enzymes that digest proteins and lipids. In contrast, the *endocrine* portion produces hormones such as insulin and glucagon, which control blood glucose levels. During the cephalic phase of digestion, even before food enters the mouth, digestive enzymes and insulin are secreted to regulate and coordinate metabolic processes.

Over millions of years of evolution, large portions of what we today call the pancreas evolved originally from just exocrine tissue. As a result of this evolution, endocrine cells form encapsulated boundaries called islets of Langerhans, that separate the endocrine and exocrine acini within the pancreas [15].

Moreover, many cell types are present within these evolved islets: the alpha (α), beta (β), and delta (δ) cells produce the vital hormones glucagon, insulin, and somatostatin, respectively. A fourth cell type, known as the pancreatic polypeptide (PP)

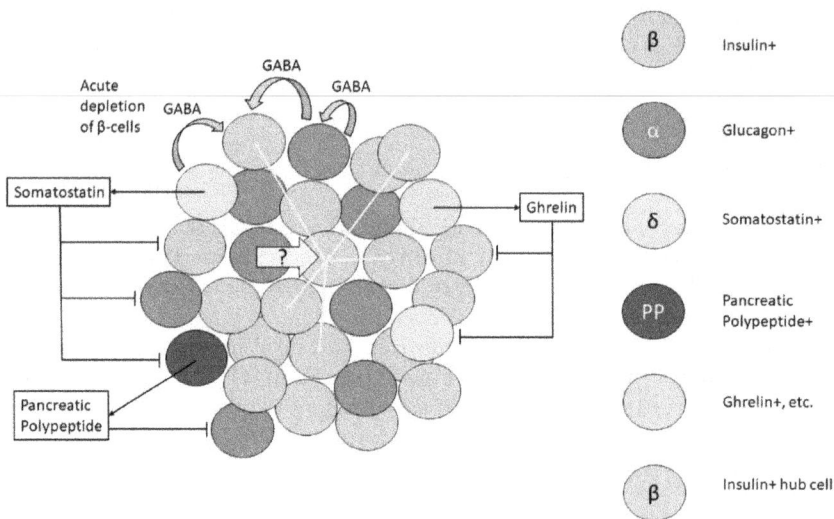

Figure 1.
Schematic diagram showing the interaction of islet cells. Evidence points to the transdifferentiation (light blue arrow) of α-cells (red) via stimulation by gamma aminobutyric acid (GABA), and δ-cells (pink) into insulin-producing β-cells (green). It is unknown whether the replacement of β-cells by α-cells is able to take the place of hub β-cells (light blue) which influence the function of other β-cells (yellow arrows). Somatostatin released from δ-cells can inhibit the release of glucagon, insulin, and pancreatic polypeptide from α-, β-, and PP cells (purple), respectively. Pancreatic polypeptide released from PP cells can inhibit the release of glucagon. Ghrelin released from ghrelin-positive islet cells (orange) can inhibit insulin and somatostatin secretion.

cell has the significant function of inhibiting glucagon release. Yet other cell types, ghrelin-positive cells, are mainly found in the gut and in the islet to inhibit insulin and somatostatin secretion and regulate the secretion of glucagon, PP, and somatostatin (**Figure 1**) [16]. Evidence in the literature, and discussed in detail by Da Silva Xavier [16], suggests transdifferentiation of α-cells via stimulation by gamma-aminobutyric acid and δ-cells into insulin-containing β (like)-cells. It is unknown whether the replenishment of β-cells from the transdifferentiation of α-cells is able to replace hub (stem) β-cells which influence the function of other β-cells (**Figure 1**) [16]. In any case, the pancreas does appear to have some reserve capacity for the regeneration of β-cells but this may be overwhelmed in DM.

In fact, the most abundant cells in the islets of Langerhans are the β-cells, comprising 55% of the cell number, and it has been suggested that they interact with other endocrine cells to influence the secretion of hormones [8]. Moreover, because the β-cells produce insulin they are the most critical pancreatic cell type involved in the etiology DM. It is therefore timely at this point to look at insulin production in more detail.

8. Insulin synthesis and secretion

Insulin production begins in the β-cells with the secretion of pre-proinsulin that is converted into proinsulin. Proinsulin is then transformed into insulin and C-peptide, which are stored in the form of secretory granules until they are triggered for release throughout the body during food ingestion. Insulin is mainly produced in response to glucose. This has been validated *in vitro*: it was found that when human islets or stem cell-derived β-cells were stimulated with glucose, they secreted insulin [17]. Other hormones, such as melatonin, estrogen, leptin, growth hormone, glucagon-like peptide 1, etc. can modulate the level of insulin secretion.

The primary signal that stimulates insulin exocytosis from granules is a process triggered by glucose (for instance, increased intake of dietary sugar) followed by a rise in intracellular calcium (Ca^{2+}) [8, 18]. Calcium influx relies on many factors such as glucose transport, metabolic enzymes, and functioning potassium ion channels. Moreover, an elegant study showed that the growth and survival effects of glucose on β-cells require activation of proteins in the insulin signaling pathway via an autocrine mechanism (**Figure 2**) [19]. This, the fact that β-cells both secrete and *respond* to insulin via autoregulation may make they especially vulnerable to epigenetic changes induced by glucose (**Figure 2**), In the model proposed by Assmann*et al.* [19], we further posit that the identified targets may be exceptionally sensitive to epigenetic dysregulation in DM (in addition to transcriptional and/or translational dysregulation) (**Figure 2**). If these targets are epigenetically misregulated they may become difficult to normalize.

Furthermore, the pancreatic endoplasmic reticulum kinase (ERK) plays a central role in regulating translational events. It regulates insulin translation through phosphorylation of eukaryotic initiation factor 2 alpha (eIF2a). ERK mutation is linked with permanent neonatal DM in humans [8]. One such example is Wolcott-Rallison syndrome (WRS), a rare autosomal recessive disease characterized by neonatal/early-onset non-autoimmune insulin-requiring DM associated with skeletal dysplasia and growth retardation [20]. Because glucose is critical in activating insulin signaling, let us look at glucose in more detail.

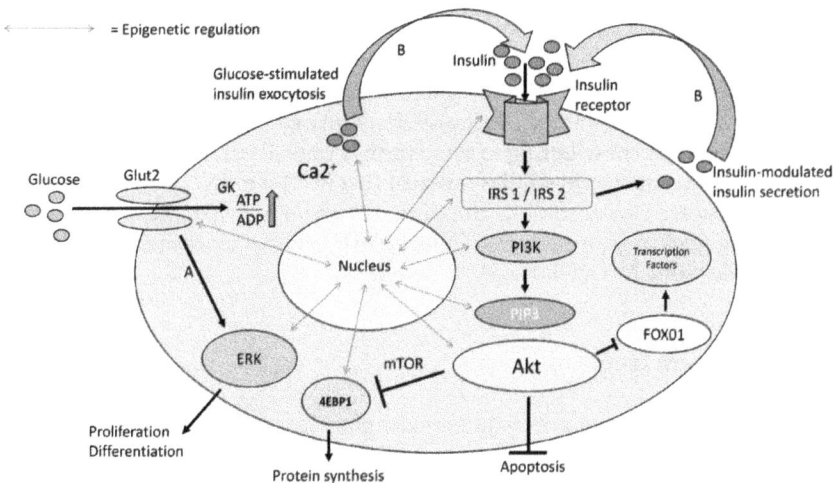

Figure 2.
Diagram of a link between glucose and insulin signaling in β-cells and indicating epigenetic effects (blue arrows). (A) Potential direct effects of glucose and/or its metabolites on proteins in the insulin/IGF-1 signaling pathway. (B) Potential indirect effects of glucose and direct effects of insulin following exocytosis of insulin. Akt, v-akt murine thymoma viral oncogene homolog; FoxO-1, forkhead box O1; GRB2, growth factor receptor-bound protein 2; PIP3, phosphatidylinositol (3,4,5)-trisphosphate; mTOR, mammalian target of rapamycin; 4EBP1, translation initiation factor 4e binding protein 1. We postulate that β-cells are especially sensitive to epigenetic perturbations (blue arrows) because unlike cells that do not produce insulin, β-cells also have an autocrine quality, in that they both produce insulin, and receive the insulin signal by receptor binding. This autoregulatory loop may be especially vulnerable to epigenetic dysregulation.

9. Glucose metabolism

Glucose activates other cell signals, such as cyclic AMP (cAMP), cyclic GMP (cGMP), inositol 1,4,5-trisphosphate (IP3), and diacylglycerol (DAG). When cAMP is produced, it then activates protein kinase A (PKA). cAMP may be the most crucial molecule that leads to insulin secretion and phosphorylation of proteins involved in insulin exocytosis via PKA. Incretin hormones also augment glucose-stimulated insulin secretion by stimulating the cAMP signaling pathway [8].

Incretin hormones are gut peptides secreted by enteroendocrine cells after feeding [21]; their function is to control the amount of insulin. In the pancreas, two kinds of incretins, glucose-dependent insulinotropic peptide (GIP) and glucagon-like peptide-1 (GLP-1), share the same behavior, but outside the pancreas, they differ. They are both rapidly deactivated by an enzyme called dipeptidyl peptidase 4 (DPP4). A decrease in incretin secretion or an increase in incretin clearance is not a pathogenic factor in DM. However, in T2D, GIP no longer modulates glucose-dependent insulin secretion, even at supraphysiological (pharmacological) plasma levels. GIP incompetence is detrimental to β-cell function, especially after eating. On the other hand, GLP-1 is still insulinotropic in T2D, which has led to the production of compounds that activate the GLP-1 receptor intending to improve insulin secretion [22].

Furthermore, glucose metabolism is critical in insulin biosynthesis because it triggers insulin gene transcription and mRNA translation. The triggering of insulin gene transcription and mRNA translation is necessary for regulating insulin biosynthesis via modification of proinsulin mRNA expression and maintaining insulin mRNA stability [8]. mRNA has a vital role in regulating and controlling gene expression and this stability is affected by how RNA-binding proteins and structural elements interact with each other.

Regulation of mRNA stability is accomplished through various reactions to developmental stimuli (e.g., nutrient levels, cytokines, hormones, and temperature shifts or to different environmental stimuli such as stresses like hypoxia, hypocalcemia, viral infection, and tissue injury). However, deregulated mRNA stability can cause mRNA accumulation contributing to some forms of neoplasia, thalassemia, and AD [23]. The results from *in vitro* studies revealed that insulin mRNA stability decreases under lower glucose concentrations and increases under high glucose conditions [8]. In the absence of glucose, insulin mRNA levels in β-cells decrease sharply, which is reversed by elevating intracellular cAMP levels.

10. Transcription factors

Another stratification of regulation besides glucose signaling and mRNA expression is at the level of transcription factors. These play a central role in regulating gene expression by binding to specific consensus sequences, or cis-elements, within promoter regions [24]. Transcription factors are proteins that can be targets of modifications when they respond to cellular stimuli. This will affect their stability, activity, intracellular distribution, and interaction with other proteins [25]. One of these stimuli is insulin resistance, which affects many organs, mainly the liver, pancreas, adipose tissue, and muscle [26, 27].

Illustrating the power of transcription factors, pancreatic acinar cells can be reprogrammed to produce, process, and secrete insulin when forced to express the transcription factors Pdx-1, MafA, and Ngn3 [28].

Cellular differentiation processes can also be negatively impacted by gene expression [8]. For example, the deletion of pancreatic and duodenal homeobox 1 (PDX-1) from postnatal islets results in phenotypic loss of the β-cells. The Pdx-1 protein is a transcription factor responsible for the development of α and β-cells [8]. A second proposed explanation for the loss of β-cell phenotype is that there is an increase in α cells in the pancreatic islets due to a lack of Pdx-1 transcriptional processes, which convert α cells to β-cells [8]. This process would produce an overall imbalance between α and β-cells; hence the β-cells would ultimately be fewer, but it is unclear whether the loss of the β-cell phenotype was due to lack of Pdx-1 or lack of α to β-cell conversion [29].

Even more starkly, mutation in both PDX-1 and PTF1A results in pancreatic agenesis [8]. PTF1A is a gene that is a component of transcription factor 1 complex (PTF1) in the pancreas and encodes a protein that functions in embryonic pancreatic development. It is crucial and determines if cells in the pancreatic buds go on either towards pancreatic organogenesis or return to duodenal fates [18, 30].

Although we see that transcription factors are powerful effectors of cellular behavior, there are deeper layers than this, namely epigenetic, which we will begin to cover in more detail.

11. Etiology of type 2 diabetes mellitus

We have covered the different types of diabetes, the cells that comprise the pancreas, glucose signaling, transcription, and some of the genes that govern pancreatic cellular behavior and differentiation, so we will now look more closely at the etiology of T2D, ultimately moving into the epigenetic layer.

The etiology of T2D is complex and multifactorial since it is affected by genetic predisposition [31] and behavioral influences, such as diet and physical activity [32]. As previously stated, T2D is often characterized by β-cell dysfunction, insulin resistance, and hyperinsulinemia [33]. These factors and symptoms depend on the disease phase and how insulin affects and regulates the bloodstream's high level of glucose [34]. Essentially, genetic, epigenetic, and non-genetic factors influence the pathogenesis of T2D [35].

Firstly, as far as genetics goes, genome-wide association studies have identified associations between single-nucleotide polymorphisms (SNPs) and disease in large case-control cohorts and family-based studies. However, although over a hundred genetic variants have been identified that are associated with T2D risk, they can explain only a modest portion of T2D heritability [36].

There also are non-genetic risk factors for T2D, such as age, physical inactivity, and energy-rich diets that result in obesity [35]. However, it is not necessary to be obese to have this type of DM. Most patients who have DM are not obese, have an incommensurable reduction of insulin secretion, and are less insulin resistant than obese individuals. It was discovered that T2D could also exist in the absence of an obese phenotype by studying non-obese rodent models [34].

There is also evidence that DM in adulthood can be caused by intrauterine or fetal malnutrition. This type of malnutrition is vital to comprehending adult DM because the genetic abnormalities and imbalances in the mother's uterus can affect the probability of her child developing DM even as an adult. Another study has shown that low birth weight leads to T2D development or insulin resistance. Moreover, factors such as the mother having DM, low birth weight of the child, and fetus malnutrition, work in a complex manner with a variety of epigenetic regulators (guided by α and β-cell-type-specific transcription factors such as Pdx1, mentioned earlier) and result in abnormal β-cell maturation and differentiation causing adult DM [37].

Thus, other explanations for T2D heritability have been proposed, including alterations in epigenetic patterns [35]. We likely need a more holistic understanding of epigenetics to obtain a complete picture of the etiology of DM, especially environmental-epigenetic interactions.

But what exactly does "epigenetic" mean? It is here that we can delve more into the molecular aspects of DNA and chromatin and how they relate to gene expression and disease etiology.

12. Role of epigenetics in diabetes mellitus

Epigenetics addresses the relationship between genes, environmental exposure, and disease development. Additionally, epigenetics concerns heritable gene expression changes *without changes in the DNA sequence* itself, affecting how cells "read" genes. Many factors affect epigenetic modifications, such as age, lifestyle, family history, and disease status. Today, three major epigenetic systems are recognized: DNA methylation, histone modifications (the most well-characterized being acetylation), and non-coding RNA (ncRNA)-associated gene silencing (**Figure 3**).

Epigenetic alterations such as DNA methylation and/or histone modifications alter the accessibility of genes to the transcriptional machinery by inducing either a relaxed/open or condensed/closed chromatin state. In general DNA methylation, principally of cytosines in gene promoters, condenses DNA and leads to gene silencing, whereas acetylation of histones opens up chromatin and is associated with

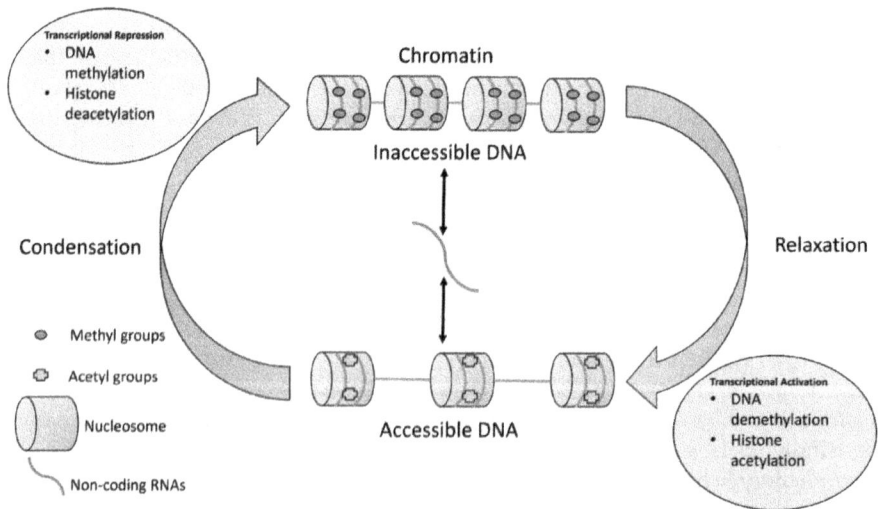

Figure 3.
Epigenetic regulation of gene expression. Epigenetic alterations such as DNA methylation and/or histone modifications alter the accessibility of genes to the transcriptional machinery by inducing either a relaxed/open or condensed/closed chromatin state. Non-coding RNAs such as miRNAs also regulate the cell phenotype by repressing or enhancing the expression of gene transcripts. Conversely, these non-coding RNAs can themselves be epigenetically regulated.

gene activation (**Figure 3**). Non-coding RNAs such as miRNAs also regulate the cell phenotype by repressing or enhancing the expression of gene transcripts (**Figure 3**). Conversely, these non-coding RNAs can themselves be epigenetically regulated. Epigenetic changes often occur during an organism's lifetime and are sometimes transmitted to the next generation [38].

Several studies suggest that epigenetics plays a vital role in the pathology of DM, especially T2D. Common T2D is likely to result from many genes interacting with different environmental factors (**Figure 4**) to produce a wide variation in the disease's clinical course [7], and as previously described for other multifactorial diseases such as hypertension [39]. In the model proposed by Arif et al. [39], epigenetic and genetic factors regulate phenotypes. Specifically, in addition to heritable Mendelian genetics, polygenic phenotypes, such as DM, are significantly affected by gene-environment interactions triggering epigenetic modifications (**Figure 4**). Indeed, previous studies have shown that epigenetic mechanisms can predispose individuals to the diabetic phenotype. Also, the altered homeostasis in T2D, such as prolonged hyperglycemia, dyslipidemia, and increased oxidative stress, could result from, and cause, epigenetic changes associated with the disease [40].

As previously stated, the main insulin-producing cells in the pancreas are the β-cells, and epigenetic modifications play a critical role in establishing and maintaining their identity and function in physiological conditions [41]. Stable β-cell function is vital to the regulation of glucose levels in the bloodstream. In the case of diabetes, epigenetic dysregulation may result in the reduction of the expression of genes essential for β-cell function, the ectopic expression of genes that are not supposed to be expressed in β-cells, and loss of genetic imprinting, leading to loss of β-cell identity [40]. Consequently, this may lead to β-cell dysfunction and impaired insulin secretion, impairing the function of the pancreas, and in turn, causing widespread sequalae and finally disease in the whole organism, Thus, a causal chain is established whereby the environment causes disease in the following sequence: environment ⟶ chromatin ⟶ genes ⟶ cells ⟶ organs ⟶ organism [42]. The model proposed by Liu *et al.* goes a long way in establishing this causal chain (**Figure 5**) [42].

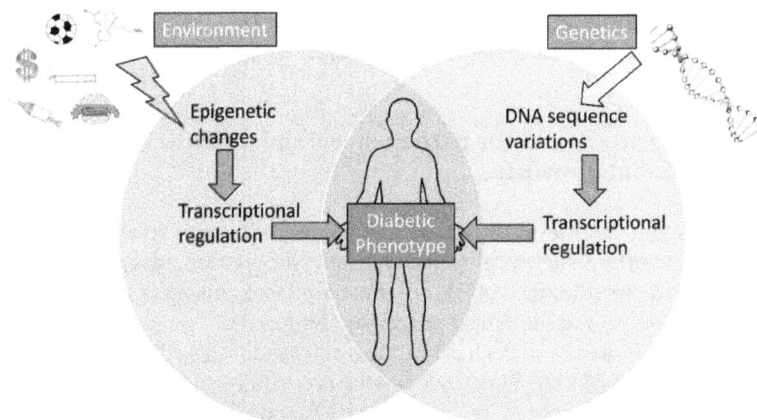

Figure 4.
Influences on the expression of phenotypes. Development of polygenic conditions, such as diabetes, depend on complex and interacting genetic and environmental pathways.

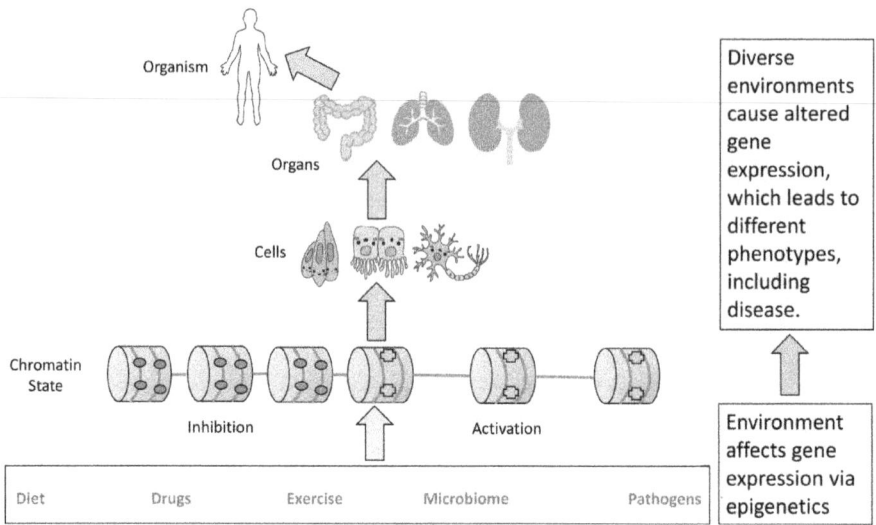

Figure 5.
A stratified view of gene-environment interactions during development and disease. Environmental effects are incorporated by epigenetic processes including chromatin remodeling to either inhibit or enhance gene expression. These effects are then manifested hierarchically in the sequence of cells to organs (i.e. pancreas) to organism. Disease etiology (for example diabetes) occurs in this hierarchical sequence.

It's important to realize that many risk factors may lead to epigenetic dysregulation by causing this initial "disruption" to chromatin, such as hyperglycemia, physical inactivity, parental obesity, mitochondrial dysfunction, aging, and an abnormal intrauterine environment. Those factors can affect the epigenome at different time points throughout the lifetime of an individual. Moreover, the epigenome can change due to environmental factors, such as diet and exercise, because of the epigenome's plasticity. As a result, the epigenome is a good target for epigenetic drugs that may be used to induce insulin secretion and treat DM [40].

Ultimately, epigenetics is vital to research DM and possible future treatments. It could be the solution to early detection and treatment, via timely detection and modification of relevant genes and reversal (normalization) of signaling pathways. But how do we identify these genes and pathways?

13. DNA methylation, and the pathogenesis and potential treatment of type 2 diabetes mellitus

As mentioned, DNA methylation and histone modifications (typically acetylation but there are others such as phosphorylation, ribosylation, ubiquitylation, sumoylation, and citrullination [43]), are the main mechanisms in which epigenetics affects cell phenotype and biological processes (**Figure 3**).

Of the two, DNA methylation has been the most well-studied by microarray. During methylation of DNA, 5-methylcytosine is created by DNA methyltransferases modifying cytosines. Most of this occurs in CpG islands in the promoter regions in multiple protein-coding genes. Methylation of cytosines at the promoter regions is associated with the repression of transcription. Repressors that bind to methylated CpG islands then initiate a cascade that results in the second primary mechanism of epigenetic regulation: namely histone modifications and the recruitment of histone deacetylases (or transferases) [27]. Repression of multiple genes could then lead to the DM phenotype. But we need to identify what these genes are.

Bansal and Pinney [44] reviewed studies where both DNA methylation and gene expression changes were reported. DNA methylation status had a strong inverse correlation with gene expression, suggesting that this may be a potential future therapeutic target. They highlighted the emerging use of genome-wide DNA methylation profiles as biomarkers to predict patients at risk of developing diabetes or specific complications of diabetes.

Indeed, developing predictive models that incorporate both genetic information *and* DNA methylation changes may be effective diagnostic approaches for *all* types of diabetes and could lead to additional innovative therapies.

For example, one study used the genome-wide Infinium 450K array and identified 1,649 CpG sites, and 853 genes that include TCF7L2, FTO, KCNQ1, IRS1, CDKN1A, and PDE7B. Significant changes in DNA methylation were found in donors that have T2D compared to controls. Also, increased DNA methylation at the promoter of CDKN1A and PDE7B was associated with decreased transcriptional activity in clonal *in vitro*, as well as impaired glucose-stimulated insulin secretion [45].

Another genome-wide study of DNA methylation using the Infinium27K array found 276 differentially-methylated CpG sites, of which 96% were hypomethylated in islets of diabetic compared to non-diabetic donors [46]. Changes in differential DNA methylation were correlated with expression changes of 34 genes assessed by microarray [46].

We are also conducting our own genome-wide methylation studies using a human *in vitro* model of diabetes based on induced pluripotent stem cell-derived β-cells.

Interestingly, bariatric surgery appears to be capable of partially *reversing* the obesity-related and diabetic epigenome [47]. The identification of potential epigenetic biomarkers predictive of the success of bariatric surgery may open new doors to personalized therapy for severe obesity and diabetes, which is cause for great optimism [47].

14. Conclusions

DNA methylation changes at differentially methylated sites or regions have been linked to metabolic diseases such as obesity and T2D. Thus, changes in the epigenome may provide an underlying molecular mechanism for the deleterious metabolic health outcomes associated with these conditions. Conversely, coordinated reversal of these changes may improve metabolic health after therapeutic intervention, and this provides optimism for the future.

Acknowledgements

Special thanks to Saudi Culture Mission (SACM) in the USA and King Saud Bin Abdulaziz University for Health Sciences in Riyadh, KSA, for their immense help and financial support to R.A.A. This publication resulted in part from research support to A.B.C. from the National Institute of Health (NIH) R25 Resource Grant (1 R25 AG047843-01) and to T.H. from the National Science Foundation [NSF IOS-1355034], Howard University College of Medicine, and the District of Columbia Center for

AIDS Research, an NIH funded program [P30AI117970], which is supported by the following NIH Co-Funding and Participating Institutes and Centers: NIAID, NCI, NICHD, NHLBI, NIDA, NIMH, NIA, NIDDK, NIMHD, NIDCR, NINR, FIC, and OAR. The content is solely the responsibility of the authors and does not necessarily represent the official views of the NIH.

Conflict of interest

The authors declare no conflict of interest.

Author details

Rasha A. Alhazzaa[1,2,3,4], Thomas Heinbockel[1] and Antonei B. Csoka[1*]

1 Department of Anatomy, Howard University College of Medicine, Washington, DC, USA

2 King Saud Bin Abdulaziz University for Health Sciences, Riyadh, Saudi Arabia

3 King Abdullah International Medical Research Center, Riyadh, Saudi Arabia

4 Ministry of the National Guard-Health Affairs, Riyadh, Saudi Arabia

*Address all correspondence to: antonei.csoka@howard.edu

IntechOpen

References

[1] Blair M. Diabetes mellitus review. Urologic Nursing. 2016;**36**:27-36

[2] Katsarou A, Gudbjörnsdottir S, Rawshani A, et al. Type 1 diabetes mellitus. Nature Reviews. Disease Primers. 2017;**3**:17016

[3] Son M-Y, Lee M-O, Jeon H, Seol B, Kim JH, Chang J-S, et al. Generation and characterization of integration-free induced pluripotent stem cells from patients with autoimmune disease. Experimental & Molecular Medicine. 2016;**48**:e232

[4] Paschou SA, Papadopoulou-Marketou N, Chrousos GP, Kanaka-Gantenbein C. On type 1 diabetes mellitus pathogenesis. Endocrine Connections. 2017;**7**:R38-R46

[5] Aggarwal I. The epidemiology, pathogenesis, and treatment of type 1 diabetes mellitus. Inquires Journal. 2015;**7**:1-2

[6] Thomas HE, Darwiche R, Corbett JA, Kay TW. Interleukin-1 plus gamma-interferon-induced pancreatic beta-cell dysfunction is mediated by beta-cell nitric oxide production. Diabetes. 2002;**51**:311-316

[7] Wolford JK, Vozarova de Courten B. Genetic basis of type 2 diabetes mellitus: Implications for therapy. Treatments in Endocrinology. 2004;**3**:257-267

[8] Fu Z, Gilbert ER, Liu D. Regulation of insulin synthesis and secretion and pancreatic Beta-cell dysfunction in diabetes. Current Diabetes Reviews. 2013;**9**:25-53

[9] Stenström G, Gottsäter A, Bakhtadze E, Berger B, Sundkvist G. Latent autoimmune diabetes in adults: Definition, prevalence, β-cell function, and treatment. Diabetes. 2005;**54**: S68-S72

[10] American Diabetes Association. Diagnosis and classification of diabetes mellitus. Diabetes Care. 2010;**33**(Suppl 1):S62-S69

[11] de la Monte SM, Wands JR. Alzheimer's disease is type 3 diabetes-evidence reviewed. Journal of Diabetes Science and Technology. 2008;**2**: 1101-1113

[12] Lee S-H, Zabolotny JM, Huang H, Lee H, Kim Y-B. Insulin in the nervous system and the mind: Functions in metabolism, memory, and mood. Molecular Metabolism. 2016;**5**:589-601

[13] Mosconi L, Mistur R, Switalski R, Tsui WH, Glodzik L, Li Y, et al. FDG-PET changes in brain glucose metabolism from normal cognition to pathologically verified Alzheimer's disease. European Journal of Nuclear Medicine and Molecular Imaging. 2009;**36**:811-822

[14] Kampmann U, Madsen LR, Skajaa GO, Iversen DS, Moeller N, Ovesen P. Gestational diabetes: A clinical update. World Journal of Diabetes. 2015;**6**:1065-1072

[15] Stanger BZ, Hebrok M. Control of cell identity in pancreas development and regeneration. Gastroenterology. 2013;**144**:1170-1179

[16] Da Silva Xavier G. The cells of the Islets of Langerhans. Journal of Clinical Medicine. 2018;**12**(7):54

[17] Pagliuca FW, Millman JR, Gürtler M, Segel M, Van Dervort A, Ryu JH, et al.

Generation of functional human pancreatic β cells in vitro. Cell. 2014;**159**:428-439

[18] Trexler AJ, Taraska JW. Regulation of insulin exocytosis by calcium-dependent protein kinase C in beta cells. Cell Calcium. 2017;**67**:1-10

[19] Assmann A, Ueki K, Winnay JN, Kadowaki T, Kulkarni RN. Glucose effects on beta-cell growth and survival require activation of insulin receptors and insulin receptor substrate 2. Molecular and Cellular Biology. 2009;**29**:3219-3228

[20] Julier C, Nicolino M. Wolcott-Rallison syndrome. Orphanet Journal of Rare Diseases. 2010;**5**:29

[21] Nauck MA, Meier JJ. Incretin hormones: Their role in health and disease. Diabetes. Obesity and Metabolism. 2018;**20**(Suppl. 1): 5-21

[22] Kim W, Egan JM. The role of incretins in glucose homeostasis and diabetes treatment. Pharmacological Reviews. 2008;**60**:470-512

[23] Guhaniyogi J, Brewer G. Regulation of mRNA stability in mammalian cells. Gene. 2001;**265**:11-23

[24] Davidson EH, Jacobs HT, Britten RJ. Very short repeats and coordinate induction of genes. Nature. 1983;**301**: 468-470

[25] Li S, Shang Y. Regulation of SRC family coactivators by post-translational modifications. Cellular Signalling. 2007;**19**:1101-1112

[26] Kahn CR. Banting Lecture. Insulin action, diabetogenes, and the cause of type II diabetes. Diabetes. 1994;**43**: 1066-1084

[27] Reaven GM. Pathophysiology of insulin resistance in human disease. Physiological Reviews. 1995;**75**:473-486

[28] Cavelti-Weder C, Zumsteg A, Li W, Zhou Q. Reprogramming of pancreatic acinar cells to functional beta cells by in vivo transduction of a polycistronic construct containing Pdx1, Ngn3, MafA in Mice. CurrProtoc Stem Cell Biology. 2017;**2**(40):4A.10.1-4A.10.12

[29] Gao T, McKenna B, Li C, Reichert M, Nguyen J, Singh T, et al. Pdx1 maintains β-cell identity and function by repressing an α-cell program. Cell Metabolism. 2014;**19**:259-271

[30] Hoang CQ, Hale MA, Azevedo-Pouly AC, Elsässer HP, Deering TG, Willet SG, et al. Transcriptional maintenance of pancreatic acinar identity, differentiation, and homeostasis by PTF1A. Molecular and Cellular Biology. 2016;**36**:3033-3047

[31] McCarthy MI. Genomics, type 2 diabetes, and obesity. The New England Journal of Medicine. 2010;**363**:2339-2350

[32] Egan B, Zierath JR. Exercise metabolism and the molecular regulation of skeletal muscle adaptation. Cell Metabolism. 2013;**17**:162-184

[33] Cerf ME. Beta cell dysfunction and insulin resistance. Frontiers in Endocrinology. 2013;**4**:37

[34] Chatzigeorgiou A, Halapas A, Kalafatakis K, Kamper E. The use of animal models in the study of diabetes mellitus. Vivo Athens Greece. 2009;**23**:245-258

[35] Davegårdh C, García-Calzón S, Bacos K, Ling C. DNA methylation in the pathogenesis of type 2 diabetes in humans. Molecular Metabolism. 2018;**14**:12-25

[36] Prasad RB, Groop L. Genetics of type 2 diabetes-pitfalls and possibilities. Genes (Basel). 2015;**6**:87-123

[37] Bernstein D, Golson ML, Kaestner KH. Epigenetic control of β-cell function and failure.Diabetes Res. Clinical Practise. 2017;**123**:24-36

[38] Nilsson E, Ling C. DNA methylation links genetics, fetal environment, and an unhealthy lifestyle to the development of type 2 diabetes. Clinical Epigenetics. 2017;**9**:105

[39] Arif M, Sadayappan S, Becker RC, et al. Epigenetic modification: A regulatory mechanism in essential hypertension. Hypertension Research. 2019;**42**:1099-1113

[40] Karachanak-Yankova S, Dimova R, Nikolova D, Nesheva D, Koprinarova M, Maslyankov S, et al. Epigenetic alterations in patients with type 2 diabetes mellitus. Balkan Journal of Medical Genetics. 2015;**18**:15-24

[41] Dayeh T, Ling C. Does epigenetic dysregulation of pancreatic islets contribute to impaired insulin secretion and type 2 diabetes? Biochemistry and Cell Biology-Biochimie et Biologie Cellulaire. 2015;**93**:511-521

[42] Liu L, Li Y, Tollefsbol TO. Gene-environment interactions and epigenetic basis of human diseases. Current Issues in Molecular Biology. 2008;**10**:25-36

[43] Moosavi A, Ardekani AM. Role of epigenetics in biology and human diseases. Iranian Biomedical Journal. 2016;**20**:246-258

[44] Bansal A, Pinney SE. DNA methylation and its role in the pathogenesis of diabetes. Pediatric Diabetes. 2017;**18**:167-177

[45] Dayeh T, Volkov P, Salö S, Hall E, Nilsson E, Olsson AH, et al. Genome-wide DNA methylation analysis of human pancreatic islets from type 2 diabetic and non-diabetic donors identifies candidate genes that influence insulin secretion. PLoS Genetics. 2014;**10**:e1004160

[46] Volkmar M, Dedeurwaerder S, Cunha DA, Ndlovu MN, Defrance M, Deplus R, et al. DNA methylation profiling identifies epigenetic dysregulation in pancreatic islets from type 2 diabetic patients. The EMBO Journal. 2012;**31**:1405-1426

[47] Izquierdo AG, Crujeiras AB. Obesity-related epigenetic changes after bariatric surgery. Frontiers in Endocrinology. 2019;**10**:232

Chapter 4

Diet-Epigenome Interactions: Epi-Drugs Modulating the Epigenetic Machinery during Cancer Prevention

Fadime Eryılmaz Pehlivan

Abstract

The roles of diet and environment on health have been known since ancient times. Cancer is both a genetic and an epigenetic disease and a complex interplay mechanism of genetic and environmental factors composed of multiple stages in which gene expression, protein and metabolite function operate synchronically. Disruption of epigenetic processes results in life-threatening diseases, in particular, cancer. Epigenetics involves altered gene expression without any change of nucleotide sequences, such as DNA methylation, histone modifications and non-coding RNAs in the regulation of genome. According to current studies, cancer is preventable with appropriate or balanced food and nutrition, in some cases. Nutrient intake is an environmental factor, and dietary components play an importent role in both cancer development and prevention. Due to epigenetic events induce changes in DNA and thus influencing over all gene expression in response to the food components, bioactive compounds and phytochemicals as potent antioxidants and cancer preventive agents have important roles in human diet. Several dietray components can alter cancer cell behavior and cancer risk by influencing key pathways and steps in carcinogenesis, including signaling, apoptosis, differentiation, or inflammation. To date, multiple biologically active food components are strongly suggested to have protective potential against cancer formation, such as methyl-group donors, fatty acids, phytochemicals, flavonoids, isothiocyanates, etc. Diet considered as a source of either carcinogens that are present in certain foods or acting in a protective manner such as vitamins, antioxidants, detoxifying substances, chelating agents etc. Thus, dietary phytochemicals as epigenetic modifiers in cancer and effects of dietary phytochemicals on gene expression and signaling pathways have been widely studied in cancer. In this chapter, current knowledge on interactions between cancer metabolism, epigenetic gene regulation, and how both processes are affected by dietary components are summerized. A comprehensive overview of natural compounds with epigenetic activity on tumorogenesis mechanisms by which natural compounds alter the cancer epigenome is provided. Studies made in epigenetics and cancer research demonstrated that genetic and epigenetic mechanisms are not separate events in cancer; they influence each other during carcinogenesis, highlighting plant-derived anticancer compounds with epigenetic mechanisms of action, and potential use in epigenetic therapy. Recent investigations involving epigenetic modulations suggest that diet rich in phytochemicals not only reduce the risk of developing cancer, but also affect the treatment outcome.

Keywords: diet, cancer, epi-drugs, epigenetic modulation, phytochemicals

1. Introduction

1.1 Epigenetics

Epigenetics is the study of the variations of genetic expression that has been referred to the heritable changes in gene expression without changes in the DNA sequence and described the interactions between the genome and the environment that leads to the formation of the phenotype [1]. Epigenetic modifications such as DNA methylation and histone modifications are able to affect gene expression mostly by interfering with transcription factors with DNA or may lead to structural rearrangement of chromatin thus promoting the expression of particular genes. These epigenetic mechanisms are those that alter the chromatin structure including DNA methylation of cytosine residues in CpG dinucleotides and post-translational histone modifications. Epigenetic regulations occur not because of differences in DNA structure, but because of chromatin alternations that modulate DNA transcription such as DNA methylation, that can mediate gene and environment interactions at the level of the genome. The mechanisms of epigenetics are thus the link between genome and phenotype [1, 2]. Epigenetic mechanisms play an important role in regulating gene expression. The main mechanisms are methylation of DNA and modifications of histones by methylation, acetylation, phosphorylation. Modifications in DNA methylation are performed by DNA methyltransferases (DNMTs) and ten-eleven translocation (TET) proteins, and by enzymes, such as histone acetyltransferases (HATs), histonedeacetylases (HDACs), histone methyltransferases (HMTs), and histone demethylases (HDMs) that regulate covalent histone modifications. In many diseases, such as cancer, the epigenetic regulatory system is disturbed [2]. MicroRNAs (miRNAs) are another epigenetic regulatory system that influences the regulation of gene expression, which are small RNA molecules, ~22 long nucleotides, that can bind to their target miRNAs and downregulate their stabilities and/or translation [3]. Recent investigations have shown the association of altered expression of noncoding RNAs in general and miRNAs in particular with epigenetic modifications [2–4], suggesting that epigenetic alterations can contribute to the carcinogenesis [3] and are considered a hallmark of cancer [4].

2. Epigenetics and cancer

The progression of cancer is driven not only by acquired genetic alterations but also epigenetic modifications [4]. Epigenetic changes have been reported during cancer development and are found in genes involved in cell differentiation, proliferation, and apoptosis [4, 5]. DNA methylation is the most extensively studied epigenetic mark which occurs on cytosines followed by guanine (CpG), in humans [4, 5]. Methylation of CpGs plays a crucial role in regulation of gene expression [5, 6], which is necessary for orchestrating key biological processes, such as cell cycle, differentiation, and genomic imprinting, where, DNA hypermethylation is found in repetitive genomic sequences to maintain these regions in a transcriptionally inactive chromatin state [4–6].

Cancer cells exhibit a global DNA hypomethylation, which causes chromosome instability leading to various mutations, loss of imprinting, activation of transposable elements disturbances in the genome, eventually, to cancer progression [5, 6]. On the other hand, a DNA hypermethylation of specific promoter regions of tumor suppressor genes leads to loss of expression of specific genes affecting pathways involved in maintenance of cellular functions, including apoptosis, DNA repair, and cell cycle, [5, 6]. Several tumor suppressor genes are silenced by promoter hypermethylation in tumors. Epigenetically mediated silencing of cyclin-dependent kinase inhibitor 2A, which is crucial for control of cell cycle has been reported in several cancers [5–7]. In addition, DNA hypermethylation-dependent silencing have been associated with the pathways regulated by microRNAs [5–7].

In cancer cells, DNA methyltransferases (DNMTs) are able to maintain DNA methylation and to de novo-methylate DNA of tumour suppressor genes [5, 6]. Recently, a new group of enzymes that induce demethylation of the DNA was found, the ten-eleven translocation (TET) enzyme family, that plays crucial roles both in tumorigenesis [5–7]. These aberrant DNA methylations are not limited to cancer cells; abnormal DNMT expressions are also linked to various diseases including cardiovascular diseases, type 2 diabetes, obesity, depression, anxiety disorder, dementia, and autism [7–9].

Gene expression is modulated by interactions between DNA methylation, histone modification, and nucleosome positioning effecting chromatin structure. Chromatin remodellers, chromatin-associated proteins, and methyl DNA binding proteins are important for structural modification of chromatin (**Figure 1**) [10].

Eukaryotic nuclei has histone proteins facilitating the dense packing of DNA and thus playing an essential role in the dynamic accessibility of DNA for transcription factors. In humans, there are two major histone families: linker histone (LH) and the core histones. The dynamic structure of chromatin allows changes in gene regulation [7–10]. The N-termini of histone proteins contain multiple lysine residues that are accessible to covalent modifications such as acetylation, methylation, phosphorylation, glycosylation, thus allowing regulation of gene transcription (**Figures 2** and **3**) [11, 12]. Aberrant expression of histone methyltransferases (HMTs), and histone demethylases (HDMs) has also been associated with cancer development [8–12].

In addition, cell cycle regulation, DNA repair mechanisms, chromosomal integrity, cellular senescence, and transcriptional activity of tumour-associated proteins such as

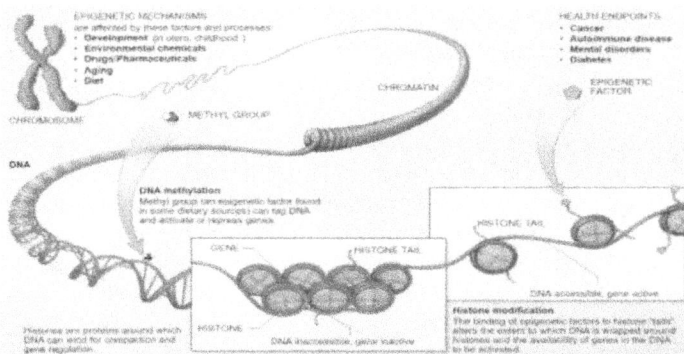

Figure 1.
Gene expression is modulated by interactions between DNA methylation, histone modification, and nucleosome positioning effecting chromatin structure [10].

DNA

Methylation

CG

Nucleosome

Histone tail

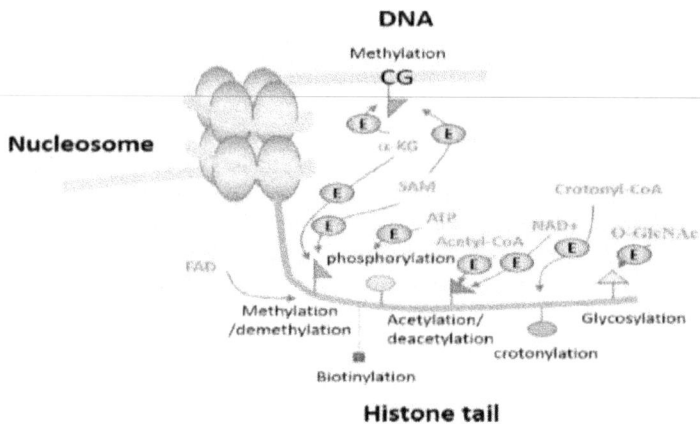

Figure 2.
Epigenetic markers on histone tails and DNA strand. Various enzymes (E) are responsible for the generation of epigenetic modification including DNA methylation/demethylation, histone acetylation/deacetylation, histone methylation/demethylation, histone biotinylation, crotonylation, phosphorylation and glycosylation [11].

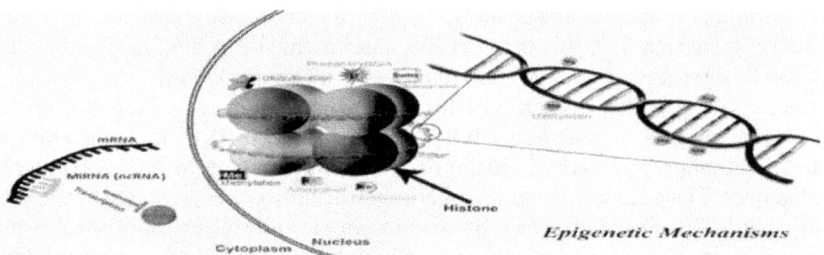

Figure 3.
Epigenetic mechanisms [12].

p53, nuclear factor kappa-lightchain- enhancer of activated B cells (NF-κB), and the FoxO family [10–14] rely on a stable cellular metabolic state. In majority of cancer cells genomic instability is found causing an increased vulnerability against DNA damaging agents [12–14]. Therefore, cancer cells might be more susceptible to exogenous compounds causing oxidative stress by production of reactive oxygen species than healthy tissues [13]. Oxidative stress plays an important role in epigenetic reprogramming of expression of tumour suppressor genes, cytokines, and oncogenes, thereby setting up a ground for carcinogenesis [13, 14].

Unlike genetic defects, epigenetic modifications are reversible and represent a promising field in therapeutic interventions [15]. Due to epigenetic aberrations occur in early stages of cancer, approaches in targeting the epigenome have been proposed as preventive and therapeutic strategies [15, 16], that aim to reverse cancer-associated epigenetic changes and restore normal gene expression. A synergistic combination of epigenetic modifying agents, including miRNAs, may provide a clinically important reversal of epigenomic cancer states.

It is known that the cause of cancer is a complex interplay mechanism of genetic and environmental factors. Dietary nutrient intake is an environmental factor and a marked variation in cancer development with the same dietary intake between individuals has been identified [17]. The effects of dietary phytochemicals on gene expression and signaling pathways have been widely studied in cancer [17, 18]. The present review aims to clarify the basic knowledge about the vital role of nutrition-related genes in cancer, focusing on the role of dietary phytochemicals as epigenetic modifiers in cancer, and summarizing the progress made in cancer chemoprevention with dietary phytochemicals.

3. Diet and cancer

Cancer is a multi stage process composed of complex stages in which gene expression, protein and metabolite function operate aberrantly [19]. Inherited mutations in genes can increase one's susceptibility for cancer; while the risk of developing cancer can be increased markedly, if there is a gene-diet interaction [19, 20]. Epigenetic functions has reversible nature which made them attractive as targets for drug development. Epigenome is continuously changing due to environmental factors such as diet, and lifestyle factors such as stress and exercise. Diet has been demonstrated to have important impact on epigenetic mechanisms [19–21]. Changes in dietary intake have been shown to affect epigenetic functions providing a significant reduction in cancer risk and also contributing to disease prevention [19–21]. In addition, revision of diet in cancer patients has shown to be resulted in changes in gene expression, that can enhance therapeutic efficacy. Diets rich in fish, fibers, fruits, vegetables, and reduction in consummation of red meat have affected the epigenome, providing therapeutic efficacy [21].

The impact of diet and environment on human health has been known since ages. Diet can either be a source of carcinogens present in certain foods or a source of protective contituents (vitamins, antioxidants, detoxifying enzyme-activating substances, etc.) [22]. Cancer initiation and progression have been linked to oxidative stress by DNA mutations, genome instability, and cell proliferation; therefore antioxidant agents could interfere with carcinogenesis. Natural herbs have been used for prevention or treatment of diverse diseases for thousands of years; depending on the presence of bioactive components in plants that makes them appropriate choices to be used as food or medicinal purposes. Plant derived bioactive components confirmed the anticancer activities of natural dietary phytochemicals which resulted in an increase in comprehension of these compounds as a biological functional agent which has a theuropetic effect on human health [22]. Epidemiological studies reported that diet rich in fruits and vegetables have cancer preventive properties and several phytochemicals originated of edible plants have defensive mechanisms that prevent the induction of carcinogenesis by scavenging free radicals and by transducting signals in response to stress factors that activate proteins associated to cellular signaling pathways [22]; thus, dietary phytochemicals are able to be a chemopreventative agent toward cancer by inflection of the cancer cell cycle, proliferation inhibition, and initiation of apoptosis [22, 23].

Common dietary compunds can act on the human genome, directly or indirectly, by altering gene expression or structure; some dietary constituents affect post translation events [23]. Acetylation of histones and non-histone proteins has been shown to affect cell metabolism and can be targeted by inhibitors of histone deacetylases (HDACs) and histone acetyl transferases (HATs) [23, 24]. Natural compounds from broccoli, garlic, curcumin speculated to have inhibitory effects of HDACs and HATs with their influence on epigenetic mechanisms for normalization of the deregulated cancer cell metabolism [23, 24]. Dietary factors can also interact with hormonal regulation such as obesity that strongly affects hormonal status such as phytoestrogens [23, 24].

Plant-derived natural bioactive compounds (phytochemicals) have acquired an important role in human diet as potent antioxidants and cancer chemopreventive agents
[23, 24]. Recently, the role of epigenetic alterations such as histone modifications, DNA
methylation, and non-coding RNAs in the regulation of genome have been addressed
(**Figure 4**) [25].

The present review outlines epigenetic mechanisms in the regulation of genome
and the role of dietary phytochemicals as epigenetic modifiers in cancer; summarizing the progress made in cancer chemoprevention with dietary phytochemicals, and
the challenges in the future.

3.1 Cancer control and prevention by diet and epigenetic approaches

Epigenetic mechanisms are known to be essential for normal development and
maintenance of adult life. Disruption of epigenetic processes results in deregulated
gene expression and leads to life-threatening diseases, in particular, cancer, which is
defined as both a genetic and epigenetic disease. Genetic and epigenetic events are
suggested to be susceptible to environmental and lifestyle factors such as radiation,
toxins, pollutants, infectious agents, and diet (**Figures 5** and **6**) [26, 27], that affect
the phenotype of cells and organisms. Diet is defined as more easily studied and
therefore better understood environmental factor in epigenetic changes [26, 27].

Cancer is known to take many years to develop from initiation to progression, as
the long period of development may represent an opportunity to use multi-functional
preventive drugs to block or reverse tumorigenesis. Unlike genetic mutations,
epigenetic alterations are potentially reversible and can be restored to their normal
state, thus one path to cancer prevention can be to target and reverse these epigenetic
defects. According to epidemiological studies there is a close link between rich diets
in bioactive compounds and the low incidence of different types of cancer; regarding
the impact of bioactive nutrients on the epigenetic mechanisms of gene expression,
such as genomic DNA methylation, altered activity and expression of DNA methyl
transferases and ten-eleven translocation enzymes, local DNA hypermethylation of
gene promoters of tumor suppressor genes or of non-coding RNAs (microRNAs and
long-noncoding RNAs), as well as global hypomethylation (**Figures 5** and **6**) [26, 27].

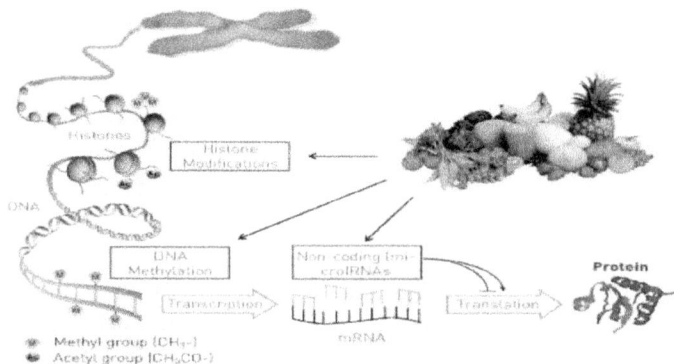

Figure 4.
*Plant-derived natural bioactive compounds (phytochemicals) have acquired an important role in human diet
as potent epigenetic modulators such as histone modifications, DNA methylation, and non-coding RNAs in the
regulation of genome (from Daniele Segnini) [25].*

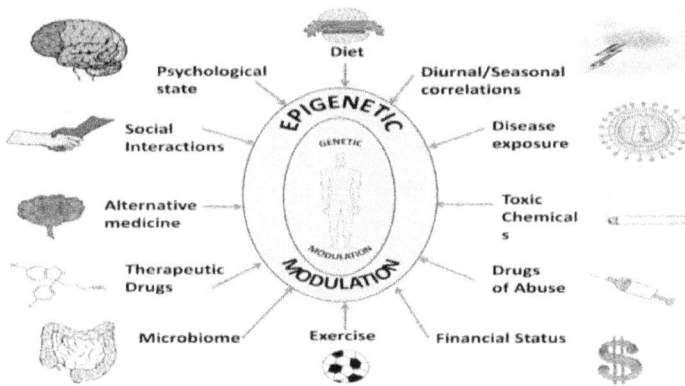

Figure 5.
Epigenetic events are suggested to be susceptible to environmental and lifestyle factors such as radiation, toxins, pollutants, infectious agents, and diet [26].

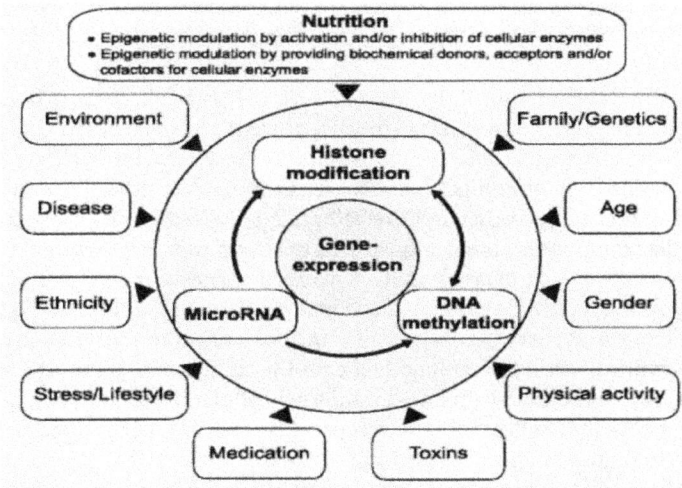

Figure 6.
Modulation and interaction of epigenetic mechanisms [27].

Dietary components play importent roles in either cancer prevention or cancer development [28–31]. Intake of certain bioactive food components such as resveratrol (grapes), polyphenol-catechins (green tea), genistein (soybean), curcumin (turmeric), sulforaphane (cruciferous vegetables), and other bioactive components such as isothiocyanate (cruciferous vegetables), apigenin (parsley), silymarin (milk thistle), cyanidins (grapes), and rosmarinic acid (rosemary) (**Figure 7**) [28] is identified to play significant roles in modulating tumor risk and development [28–31].

Despite the investigations that epigenetic changes are heritable in somatic cells and epimutations are rare in healthy tissues, it is of interest to note that epigenetic modulations are potentially reversible. Depending on this property targeting epigenetic mechanisms have been a promising approach for cancer prevention [32]. Interestingly, altered diet is found to have transgenerational effects [33]. In a study done by Heijmans *et al.* [33] pregnant mothers during the Dutch Hunger Winter of

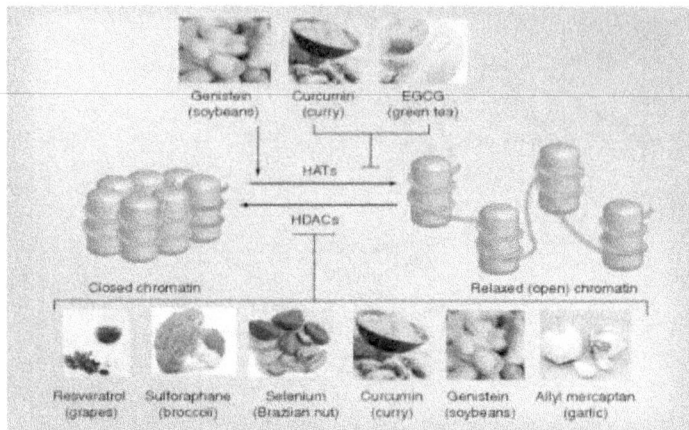

Figure 7.
Certain bioactive food components such as resveratrol (grapes), polyphenol-catechins (green tea), genistein (soybean), curcumin (turmeric), sulforaphane (cruciferous vegetables), and other bioactive components such as isothiocyanate (cruciferous vegetables) playing significant roles in modulating tumor risk and development [28].

1944 to 1945; methylation profiles of the mothers' offspring six decades later are followed and compared them with the profiles of their unexposed, same-sex siblings. The data indicated hypomethylation of insulin-like growth factor 2 (*IGF2*) and hypermethylation of interleukin-10 (*IL-10*), *LEP*, *ABCA1*, and *MEGF*; indicating the significance of diet components in the development of diseases including cancer [34].

Several diet components are demonstrated to alter tumor cell behavior and cancer risk by influencing key pathways and steps in carcinogenesis, including inflammation, cell signaling, cell cycle control, hormonal regulation, apoptosis, differentiation, and carcinogen metabolism [31–34]. While, antioxidant compunds such as polyphenols and resveratrol, are known to modulate proliferating cell nuclear antigen, *p21*, and *p27* [34, 35]; and indole-3-carbinol inhibiting cellular proliferation in human breast cancer cells [34, 35]; xenobiotic compounds, such as tobacco-specific carcinogens known to induce lung cancer [36].

Epigenetic modifications such as DNA methylation, histone modifications, chromatin remodeling, and non-coding RNAs are the most common epigenetic mechanisms. Dietary agents such as sulforaphane (SFN) found in cruciferous plants and epigallocatechin-3-gallate (EGCG) in green tea are demonstrated to exhibit various epigenetic mechanisms such as histone modifications via histone deacetylase (HDAC), histone acetyltransferase (HAT) inhibition, DNA methyltransferase (DNMT) inhibition, or noncoding RNA expression [37, 38]. These phytochemicals are shown to have an enhanced effect on epigenetic changes, which play a crucial role in cancer prevention [37, 38]. Meanwhile, restriction of glucose has been suggested to decrease the incidence of cancer and diabetes. Diet rich in compounds such as SFN and EGCG are reported to modulate the epigenome positively and lead to many health benefits; while reducing glucose in the diet is conferred to reduced cancer incidence [37, 38]. As a result, due to change in lifestyle and food habits, people can reduce risk of diet-related diseases and cancers. This review is focused on the phytochemicals that can affect various epigenetic modifications such as DNA methylation and histone modifications as well as regulation of non-coding miRNAs expression for treatment and prevention of various types of cancer.

3.2 Dietary compounds as epigenetic modulating agents in cancer

Drugs targeting epigenetic processes are called "epi-drugs", which are mostly plant-derived compounds that work through epigenetic mechanisms such as polyphenols, alkaloids, organosulfur compounds, and terpenoids [39]. Epigenetic mechanisms such as DNA methylation and posttranslational histone modifications regulate expression of various genes of changes in the DNA sequence, that play important roles in controlling cellular functions, including the cell cycle, signal transduction and immunoresponses [40]. On the other hand, epigenetic aberrations are associated with proliferation of cancer cells and oncogenesis, that these epigenetic alterations have been identified in many human cancer cells [40, 41]. This review focuses on the plant-derived anticancer drugs with epigenetic mechanisms of action, and their potential use in epigenetic therapy.

3.2.1 Therapeutic potential of polyphenols on DNA methylation

Plant-derived flavonoids as a therapeutic agents for cancer, attributed to their ability for epigenetic regulation of cancer pathogenesis [42]. The epigenetic mechanisms of various classes of flavonoids including flavonols, flavones, isoflavones, flavanones, flavan-3-ols, and anthocyanidins, such as cyanidin, delphinidin, and pelargonidin, are demonstrated [43]. These phytochemicals are mainly contained in fruits, vegetables and seeds, as well as in dietary supplements; that act as powerful antioxidants and anti-carcinogen agents; such as curcumin, catechins, genistein, quercetin and resveratrol.

As known, epigenetic modifications of chromatin are reversible and inherited, so they represent promising targets for the development of novel drugs targeting the epigenome which can contribute to amelioration of conventional therapies in cancer [44]. It has been reported that a diet rich in phytochemicals may act through epigenetic mechanisms such as modulation of DNMTs and HDACs activities that can significantly reduce the risk of cancer development by regulating the expression of oncogenes and tumor suppressor genes [45]. Cancer treatments are involved using chemo-radio therapeutic agents, kinase inhibitors, antibodies as well as certain compounds that stimulate the immune system, generally. Meanwhile, demethylating drugs modified gene expressions by reversing the aberrant epigenetic alterations acquired during tumorigenesis [44, 45]. In this context, polyphenolic flavonoid compounds may represent an alternative therapeutic option for cancer treatment.

Flavonoids are natural phenolic molecules that form a large group of secondary plant metabolites with important biological activities; subgroups of flavonoids are: flavonols such as quercetin, kaempferol, and myricetin; that are found in onions, curly, broccoli. The flavanones as hesperetin and naringenin that are found in grapefruit, oranges, and lemons. Isoflavonoids including daidzein and genistein are found in leguminous. The flavones as apigenin and luteolin that are present in cereals. The flavanols as catechin are found in green tea and chocolate, and the anthocyanins including cyanidin, delphinidin, malvidin, pelargonidin, peonidin, and petunidin are present in berries, pears, apples, grapes and peaches [46]. The biological effects of flavonoids have been linked to their antioxidative activities, that these compounds inhibit cell proliferation, induce cytotoxicity, suppression of angiogenesis; and situmulation of apoptosis, in cancer (**Figure 8**) [27]; displaying diverse properties affecting epigenetic mechanisms such as modulation of the DNA methylation and histone acetylation [23–25].

Phytochemicals and other bioactive dietary compounds are reported to restore global and gene-specific promoter DNA methylation patterns by reactivating

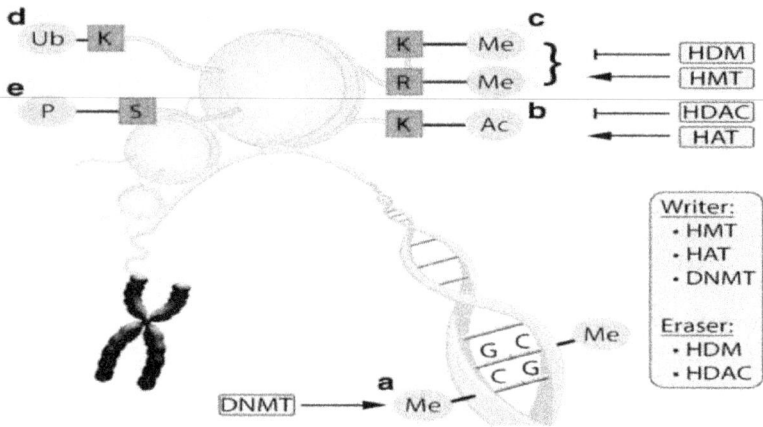

Figure 8.
Epigenetic machinery that is included the regulation of DNMTs and HDACs activities [27] (from Jan Frank).

DNA methyltransferases or by providing the provision of methyl groups [47]. This review focuses on the impact of modified DNA methylation pattern on early carcinogenesis and summarizes the effects/mechanism of phytochemical interventions on this type of epigenetic alterations. Recent investigations reported that flavonoids blocked the development and progression of tumors by targeting key signaling transducers resulting in the restoration of tumor suppressor genes, and inhibition of oncogenes expression [44–47] by modulating epigenetic machinery that is included the regulation of DNMTs and HDACs activities (**Figure 8**) [27].

Depending on epidemiological studies, dietary flavonoid intake is strongly suggested to reduce the risk of numerous cancer entities. According to current studies, cancer is preventable with appropriate or balanced diet and avoidance of obesity [48–50], in some cases. Multiple biologically active food components are strongly suggested to have protective potential against cancer formation, examples are methyl-group donors such as phytochemicals, flavonoids, isothiocyanates, allyl compounds, selenium and fatty acids [49–51].

3.2.1.1 Epigenetic effects of curcumin in cancer prevention

The yellow pigment curcumin (diferuloyl methane), a polyphenolic compound derived from turmeric (*Curcuma longa* Linn), a major ingredient of the spice curry, possessing remarkable antioxidant properties and has been studied for its potential anti-anticancer effects [52]. It has a broad spectrum of activities and acts on signaling pathways, particularly NF-κB signaling; has been shown to induce apoptosis and block invasion, metastasis, and angiogenesis for all major tumor entities [52]. It has been reported to modulate epigenetic changes in cancer cells, and has been shown to be a DNA hypomethylation agent in colon, prostate, and breast cancer, thus serving as a chemopreventive agent [53], other epigenetic studies include histone acetylation/deacetylation, and histone methylation/demethylation [52, 53]. Curcumin is first identified as an inhibitor of HAT activity; as a specific inhibitor of p300, also identified as inhibitor of acetylation of the tumor suppressor protein p53 as a non-histone protein target of p300 [52, 53], considering acetylation of p53 to be essential for p53-dependent growth arrest and apoptosis; curcumin and its derivatives have also been identified as potent modulators of miRNAs [53].

3.2.1.2 Epigenetic effects of isothiocyanates in cancer prevention

Organosulfur compounds, isothiocyanates (ITCs), are the most investigated glucosinolate-derived bioactive diet components. The chemopreventive properties of ITCs on cancer, are well demonstrated [18, 38, 54]. Although the anticancer effects of ITCs, little attention has focused on their ability about the epigenetic processes that lead to epigenetic changes in cancer. Regular intake of organosulfur compounds is reported to protect cardiovascular health [18, 38, 54], besides prevent carcinogenesis stimulated by N-nitrosodiethylamine [18, 38, 54].

3.2.1.3 Epigenetic effects of green tea polyphenols in cancer prevention

Green tea polyphenols constitute a mixture of flavan-3-ols containing a catechol moiety. Biochemical compounds from green tea such as (−)-epigallocatechin gallate have been demonstrated to alter DNA methyltransferase activity in studies of various cancer cells. Mouse model studies have confirmed the inhibitory effect of (−)-epigallocatechin gallate on DNA methylation [27, 55].

3.2.1.4 Epigenetic effects of quercetin in cancer prevention

Quercetin is reported to have a broad spectrum of cancer-preventive activities: acting as an antioxidant and modulating enzymes and signaling cascades involved in detoxification, inflammation, proliferation, apoptosis, angiogenesis, autophagy, immune defense, and senescence; besides it has been suggested to have potential to inhibit DNMT activity *in vitro*, associated with p16 up-regulation at the mRNA and protein level and inhibition of cell proliferation [27, 56].

3.2.1.5 Epigenetic effects of resveratrol in cancer prevention

Resveratrol is a plant-derived stilbene derivative found in fruits, especially in the skin of red grapes [45]. It has been reported to have a broad spectrum of health-beneficial effects, including antioxidant, cardioprotective, and antitumor activities, which have mechanistically been linked to effects on cell signaling related to cell survival, apoptosis, inflammation and tumor angiogenesis [45, 46]. Resveratrol was shown to prevent carcinogenesis in animal models for various cancer typesi, and reduced xenograft growth of various tumor cell lines. For example, activation of the aryl hydrocarbon receptor (AhR) has been shown to lead to epigenetic silencing of the DNA repair gene BRCA1 in breast cancer [57–59].

3.2.1.6 Epigenetic effects of anacardic acid in cancer prevention

A component of cashew nut shell liquid, anacardic acid (6-nonadecyl salicylic acid), is identified as a natural-product inhibitor of the HAT enzyme which is involved in the activation of key enzymes of DNA damage response, which is also found to inhibit p300-mediatedacetylation of the p65 subunit of NF-κB (nuclear factor "kappa light-chain enhancer" of activated B cells) as a non-histone substrate of HATs, and inhibited NF-κB-mediated signaling involved in inflammation, cell survival, proliferation, and invasion [60–62].

3.2.1.7 Epigenetic effects of gallic acid in cancer prevention

Gallic acid (3,4,5-trihydroxybenzoic acid), having high antioxidant activity is found in various fruit, tea and coffee, witch hazel, sumach, oak bark, walnuts, berries and other plants, as free tannins and as part of hydrolyzable tannins (gallotannins) [63]; has been shown to reduce oxidative DNA damage and to induce apoptosis in cancer cells [64]. It is identified as a specific inhibitör of HAT activity *in vitro*, and finally reduced NF-κB activation and expression of anti-apoptotic genes in response to pro-inflammatory stimuli [64].

3.2.1.8 Epigenetic effects of delphinidin in cancer prevention

Fruit, particularly blueberries contain anthocyanidins that have high antioxidant potential; possessing antiproliferative activity, inducing apoptosis and cell differentiation, and inhibiting angiogenesis and invasiveness, contributing to their high chemopreventive potential [60, 61, 65]. Overall, anthocyanidins have been shown to prevent cancer, and delphinidin has been identified as a HAT inhibitor [65]. HAT-mediated acetylation of histones and non-histone proteins seems to play an important role and; as gallic acid, delphinidin is proved to reduce pro-inflammatory signaling by preventing acetylation of the NF-κB [65], contributing to the anti-inflammatory activity of chemopreventive polyphenols [65].

3.2.1.9 Epigenetic effects of flavolignan silymarin in cancer prevention

Milk thistle (*Silybum marianum*) is used to protect liver against various diseases, and poisions. Silymarin is derived from milk-thistle seeds contains at least seven flavolignans and additional components. The most abundant compound is silybinin (or silibinin), existing as isomers, silybin A and B. Cancer-preventive potential of milk-thistle has been attributed to the inhibition of cell growth, angiogenesis, tumor invasion, metastasis, and inflammation [66, 67]. It is reported that silybinin treatment reduced the growth of human liver cancer xenografts through induction of apoptosis, and this was associated with an increase in histone H3 and H4 acetylation [68].

3.2.1.10 Epigenetic effects of genistein and soy isoflavones in cancer prevention

Isoflavones (genistein and daidzein) are a class of flavonoids found in plants of the *Fabaceae* family abundantly, and characterized by phytoestrogenic properties. They are contained in high amounts in soybean (*Glycine max* L.) and are enriched in soy products. Epidemiological studies indicates an inverse correlation between a traditional soy-rich, low-fat Asian diet and the risk of developing breast and prostate cancer [69, 70]. As soy isoflavones and phytoestrogens bind to the estrogen receptor and modulate ER signaling; genistein has been shown to affect several additional chemopreventive mechanisms, including inhibition of oxidative stress, activation of carcinogens, cell signaling, angiogenesis, modulation of cell-cycle regulation, induction of apoptosis and inhibition of inflammation [71]. Recent investigations of a growth-promoting effect of genistein in ER-positive breast-cancer cell lines and xenograft models have indicated a potential risk of genistein for human health [72]; while another recent review does not support these concerns that genistein is tested in various clinical trials for the treatment and prevention of prostate, bladder, kidney, breast, and endometrial cancer [73]. Nutrients are classified that supply methyl groups have been shown to have a protective effect in various cancer types, such as curcumin, isothiocyanates, green tea polyphenolics, quercetin, resveratrol, anacardic acid, gallic acid, delphinidin, silymarin, silybidin, and genistein that are found in various food components and medicinal plants are summerized in this chapter (**Figure 9**) [74].

These were the best known bioactive food compounds; besides these dietary components folic acid, alliin and allicin in garlic, omega 3 fatty acids, pigments such as licopene, carotenoids and anthocyanins, multivitamins such as vitamine A, C, E, vitamine B12 moreover, minerals such as zinc and selenium are the examples of nutrients that have a proven role in cancer prevention through an epigenetic mechanism [59–61, 74–77]; that substantially take part in prevention of various cancer types such as oral, breast, skin, esophageal, colorectal, prostate, pancreatic and lung cancers (**Figure 10**) [74].

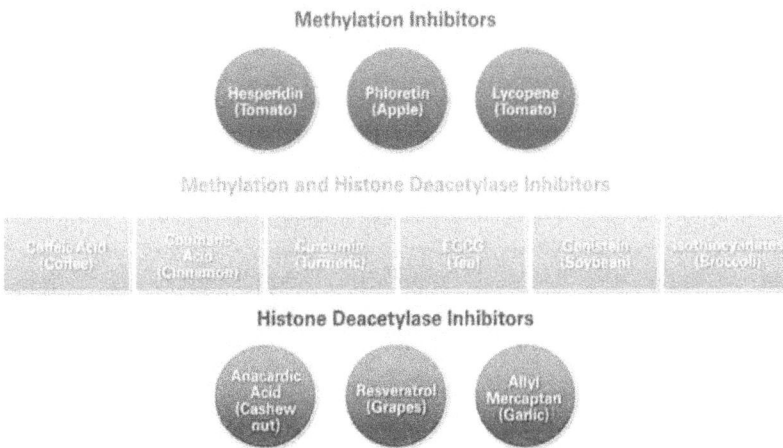

Figure 9.
Natural food components with epigenome altering properties [74].

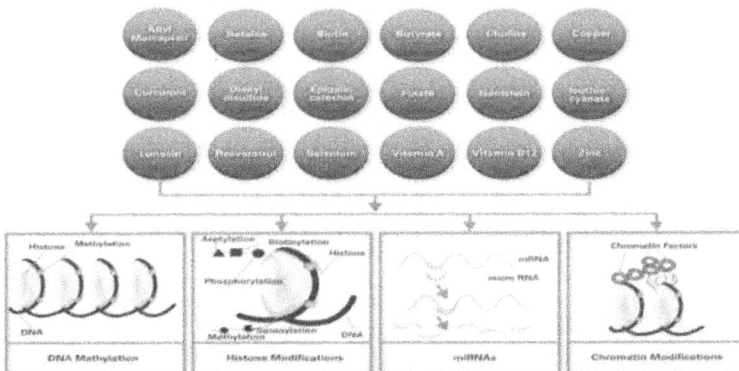

Figure 10.
Dietary components and their interaction with epigenetic regulation [74].

4. Conclusion

Epigenetic modifications is observed to perform a significant role in disease occurrence and pathogenesis. DNA methylation and chromatin remodeling are the most common epigenetic mechanisms, as described as a phenomenon of modifications in gene expression caused by heritable, but reversible, alterations in the chromatin structure, DNA methylation, and post-transcriptional effects of small noncoding microRNAs (miRNAs), without changes in the DNA sequence. The relationship between epigenome, epigenetic mechanisms and gene expression form a complicated feedback network that regulates and organizes cellular functioning at the molecular level; when this regulatory circuit is disrupted by internal or external factors, normal physiological functions are affected, leading to tumor initiation process [59]. Epigenetic mechanisms represent novel targets for natural products in prevention and treatment of cancer and other diseases. The influence of various classes of diet phytochemicals on the enzymatic activities of enzymes involved in epigenetic gene regulation; such as DNA and histone methyltransferases (DNMTs and HMTs), histone acetyltransferases (HATs), histone deacetylases (HDACs), and histone demethylases (HDMs) are also emphasized.

As a conclusion, the present review provided an overview of the most frequent epigenetic alterations in cancers, then described the most studied dietary phytochemicals and their potential use in the reversion of cancer hallmarks through epigenetic mechanisms, and finally discussed their potential use as an alternative strategy for cancer therapy. Above all, this review focused on modulation of epigenetic activities by epi-drugs that will allow the discovery of novel biomarkers for cancer prevention, as a potential alternative therapeutic approach in cancer, summarizing the progress made in cancer chemoprevention with dietary phytochemicals, and challenges in the future.

Author details

Fadime Eryılmaz Pehlivan
Istanbul University Faculty of Science Department of Biology Section of Botany, Fatih-Istanbul, Turkey

*Address all correspondence to: eryilmazfadime@gmail.com

IntechOpen

References

[1] Loscalzo J. and Handy DE. Epigenetic modifications: basic mechanisms and role in cardiovascular disease (2013 Grover Conference series). Pulmonary Circulation 2014;**4**(2):169-174.

[2] Tchurikov NA. Molecular Mechanisms of Epigenetics. Biochemistry (Moscow); 2005; **70**(4): 406-23

[3] Tahiliani M, Koh KP, Shen Y, Pastor WA, Bandukwala H, Brudno Y, Agarwal S, Iyer LM, Liu DR, Aravind L, Rao A. Conversion of 5-methylcytosine to 5-hydroxymethylcytosine in mammalian DNA by MLL partner TET1. Science 2009; 324; 930-935

[4] Fraga MF, Esteller M. Towards the human cancer epigenome: a first draft of histone modifications. Cell Cycle. 2005;4:1377-81

[5] Moore LD, Le T, Fan G. DNA methylation and its basic function. Neuropsychopharmacology. 2013; 38(1):23-38

[6] Miller JL, Grant PA. The role of DNA methylation and histone modifications in transcriptional regulation in humans. Subcell Biochem. 2013;61:289-317

[7] Fetahu IS, Höbaus J, Kállay E. Vitamin D and the epigenome. Front Physiol. 2014;5:164.

[8] Saito K., Nishida KM, Mori T, Kawamura Y, Miyoshi K, Nagami T, Siomi H, Siomi MC. Specific association of Piwi with rasiRNAs derived from retrotransposon and heterochromatic regions in the Drosophila genome. Genes Dev. 2006; **20**(16): 2214--2222.

[9] Stepanić V, Kujundzic RN, & Trošelj KG. Epigenome, Cancer Prevention and Flavonoids and Curcumin. In: Epigenetics and Epigenomics. Christopher J. Payne (Ed.) InTech Open; 2014; doi.10.5772/58247

[10] Alain LF. Genetics, Epigenetics and Cancer. Canc Therapy & Oncol Int J. 2017; **4**(2): 555634. doi: 10.19080/CTOIJ.2017.04.555634

[11] Zhao Z., Wang L. & Di LJ. Compartmentation of Metabolites in Regulating Epigenomes of Cancer. Mol Med 2016; **22:** 349-360 https://doi.org/10.2119/molmed.2016.00051

[12] Hodjat M, Rahmani S, Khan F. Niaz K, Navaei–Nigjeh M, Nejad SM, Abdollahi, M. Environmental toxicants, incidence of degenerative diseases, and therapies from the epigenetic point of view. Arch Toxicol. 2017; 91, 2577-2597://doi.org/10.1007/s00204-017-1979-9

[13] Qian Y, Chen X. Senescence regulation by the p53 protein family. Methods Mol Biol. 2013; 965:37-61. doi: 10.1007/978-1-62703-239-1_3.

[14] Rosângela FF de Araújo, Danyelly Bruneska G. Martins and Maria Amélia C.S.M. Borba. Oxidative Stress and Disease. InTechOpen. 2016; 12-21

[15] Kelly TK, De Carvalho DD, Jones PA. Epigenetic modifications as therapeutic targets. Nat Biotechnol. 2010; **28**(10):1069-1078. doi:10.1038/nbt.1678

[16] Baylin SB, Jones PA. Epigenetic Determinants of Cancer. Cold Spring Harb Perspect Biol. 2016; **8**(9):a019505. doi:10.1101/cshperspect.a019505

[17] Elsamanoudy AZ, Mohamed Neamat-Allah MA, Hisham Mohammad FA, Hassanien M, Nada HA.

The role of nutrition related genes and nutrigenetics in understanding the pathogenesis of cancer. J Microsc Ultrastruct. 2016; **4**(3):115-122.

[18] Hardy TM, Tollefsbol TO. Epigenetic diet: impact on the epigenome and cancer. Epigenomics. 2011; **3**(4):503-518. doi:10.2217/epi.11.71

[19] Sharma S, Kelly, TK, Jones PA. Epigenetics in cancer. Carcinogenesis. 2010; **31**(1), 27-36.

[20] Tiffon C. The Impact of Nutrition and Environmental Epigenetics on Human Health and Disease. Int J Mol Sci. 2018; **19**(11):3425.

[21] Lundstrom K. Epigenetics: Diet and Cancer. Austin J Genet Genomic Res. 2016; **3**(1): 1020.

[22] Thakur VS, Deb G, Babcook MA, Gupta S. Plant phytochemicals as epigenetic modulators: role in cancer chemoprevention. AAPS J. 2014; **16**(1):151-163. doi:10.1208/s12248-013-9548-5

[23] Kotecha R, Takami A, Espinoza JL. Dietary phytochemicals and cancer chemoprevention: a review of the clinical evidence. Oncotarget. 2016; 7(32):52517-52529. doi:10.18632/oncotarget.9593

[24] Knackstedt RW, Moseley VR, Wargovich MJ. Epigenetic mechanisms underlying diet-sourced compounds in the prevention and treatment of gastrointestinal cancer. Anticancer Agents Med Chem. 2012; **12**(10):1203-1210. doi:10.2174/187152012803833053

[25] https://www.danielesegnini.it/epigenetica-e-nutrizione/ Daniele Segnini, Epigenetica e nutrizione

[26] Kanherkar R, Bhatia-Dey N, and Csoka AB. Epigenetics across the human lifespan. Front. Cell Develop Bio. 2014; 2(49): 1-19.

[27] Busch C, Burkard M, Leischner C, Lauer UM, Frank J, Venturelli S. Epigenetic activities of flavonoids in the prevention and treatment of cancer. Clin Epigenetics. 2015; **10**;7(1):64. doi: 10.1186/s13148-015-0095-z.

[28] Hardy TM, Tollefsbol TO. Epigenetic diet: impact on the epigenome and cancer. Epigenomics. 2011; **3**(4):503-18. doi: 10.2217/epi.11.71.

[29] Kanwal R, Gupta S. Epigenetic modifications in cancer. Clin Genet. 2012; **81**(4):303-311. doi:10.1111/j.1399-0004.2011.01809.x

[30] Zhang Y, Kutateladze TG. Diet and the epigenome. Nat Commun. 2018; **9**(1):3375. doi:10.1038/s41467-018-05778-1

[31] Shankar E, Kanwal R, Candamo M, Gupta S. Dietary phytochemicals as epigenetic modifiers in cancer: Promise and challenges. Seminars in Cancer Biology. 2016; **40**-41:82-99.

[32] Fardi M, Solali S, Farshdousti Hagh M. Epigenetic mechanisms as a new approach in cancer treatment: An updated review. Genes Dis. 2018; **5**(4):304-311. doi:10.1016/j.gendis.2018.06.003

[33] Heijmans BT, Tobi EW, Stein AD, Putter H, Blauw GJ, Susser ES, Slagboom PE, Lumey LH. Persistent epigenetic differences associated with prenatal exposure to famine in humans. Proc Natl Acad Sci U S A. 2008; **105**(44):17046-17049. doi:10.1073/pnas.0806560105

[34] Verma M. Cancer control and prevention by nutrition and epigenetic approaches. Antioxid Redox Signal. 2012; **17**(2):355-364. doi:10.1089/ars.2011.4388

[35] Losada-Echeberría M, Herranz-López M, Micol V, Barrajón-Catalán E. Polyphenols as Promising Drugs against Main Breast Cancer Signatures. Antioxidants (Basel). 2017; **6**(4):88. doi:10.3390/antiox6040088

[36] Tan XL, Spivack SD. Dietary chemoprevention strategies for induction of phase II xenobiotic-metabolizing enzymes in lung carcinogenesis: A review. Lung Cancer. 2009; **65**(2):129-137. doi:10.1016/j.lungcan.2009.01.002

[37] Gao Y, Tollefsbol TO. Impact of Epigenetic Dietary Components on Cancer through Histone Modifications. Curr Med Chem. 2015; **22**(17):2051-2064. doi:10.2174/09298673226661504 20102641

[38] Daniel M, Tollefsbol TO. Epigenetic linkage of aging, cancer and nutrition. J Exp Biol. 2015; **218** (1):59-70. doi:10.1242/jeb.107110

[39] Schneider-Stock R, Ghantous A, Bajbouj K, Saikali M, Darwiche N. Epigenetic mechanisms of plant-derived anticancer drugs. Frontiers in Bioscience (Landmark Edition). 2012;17:129-173. DOI: 10.2741/3919.

[40] Moosavi A, Motevalizadeh Ardekani A. Role of Epigenetics in Biology and Human Diseases. Iran Biomed J. 2016; **20**(5):246-258. doi:10.22045/ibj.2016.01

[41] Ducasse M, Brown MA. Epigenetic aberrations and cancer. Mol Cancer. 2006; 5:60. doi:10.1186/1476-4598-5-60

[42] Guo Y, Su ZY, Kong AN. Current Perspectives on Epigenetic Modifications by Dietary Chemopreventive and Herbal Phytochemicals. Curr Pharmacol Rep. 2015; **1**(4):245-257. doi:10.1007/s40495-015-0023-0

[43] Khan H, Belwal T, Efferth T, Farooqi AA, Sanches-Silva A, Vacca RA, Nabavi SF, Khan F, Prasad Devkota H, Barreca D, Sureda A, Tejada S, Dacrema M, Daglia M, Suntar İ, Xu S, Ullah H, Battino M, Giampieri F, Nabavi SM. Targeting epigenetics in cancer: therapeutic potential of flavonoids. Crit Rev Food Sci Nutr. 2020; 1-24. doi: 10.1080/10408398.2020.1763910.

[44] Carlos-Reyes Á, López-González JS, Meneses-Flores M, et al. Dietary Compounds as Epigenetic Modulating Agents in Cancer. Front Genet. 2019;10:79. doi:10.3389/fgene.2019.00079

[45] Pop S, Enciu AM, Tarcomnicu I, Gille E, Tanase C. Phytochemicals in cancer prevention: modulating epigenetic alterations of DNA methylation. Phytochem Rev. 2019; **18**, 1005-1024.

[46] Moga MA, Dimienescu OG, Arvatescu CA, Mironescu A, Dracea L, Ples L. The Role of Natural Polyphenols in the Prevention and Treatment of Cervical Cancer-An Overview. Molecules. 2016; **21**(8):1055. doi:10.3390/molecules21081055

[47] Lanzotti V, Carteni F. Drugs based on natural compounds: recent achievements and future perspectives. Phytochem Rev. 2019; **18**, 967-969.

[48] Rodríguez-García C, Sánchez-Quesada C, J Gaforio J. Dietary Flavonoids as Cancer Chemopreventive Agents: An Updated Review of Human Studies. Antioxidants (Basel). 2019; **8**(5):137. doi:10.3390/antiox8050137

[49] Ong TP, Moreno FS, Ross SA. Targeting the epigenome with bioactive food components for cancer prevention. J Nutrigenet Nutrigenomics. 2011; **4**(5):275-292. doi:10.1159/000334585

[50] Wang J, Jiang YF. Natural compounds as anticancer agents: Experimental evidence. World Journal of Experimental Medicine. 2012; **2**(3):45-57. doi: 10.5493/wjem.v2.i3.45.

[51] Niedzwiecki A, Roomi MW, Kalinovsky T, Rath M. Anticancer Efficacy of Polyphenols and Their Combinations. Nutrients. 2016; **8**(9):552. doi:10.3390/nu8090552

[52] Sharifi-Rad J, Rayess YE, Rizk AA, Sadaka C, Zgheib R, Zam W, Sestito S, Rapposelli S, Neffe-Skocińska K, Zielińska D, Salehi B, Setzer WN, Dosoky NS, Taheri Y, El Beyrouthy M, Martorell M, Ostrander EA, Suleria HAR, Cho WC, Maroyi A, Martins N. Turmeric and Its Major Compound Curcumin on Health: Bioactive Effects and Safety Profiles for Food, Pharmaceutical, Biotechnological and Medicinal Applications. Front Pharmacol. 2020; **11**:01021. doi:10.3389/fphar.2020.01021

[53] Hassan FU, Rehman MS, Khan MS, Ali MA, Javed A, Nawaz A, Yang C. Curcumin as an Alternative Epigenetic Modulator: Mechanism of Action and Potential Effects. Front Genet. 2019; 10:514. doi:10.3389/fgene.2019.00514

[54] Montgomery M, Srinivasan A. Epigenetic Gene Regulation by Dietary Compounds in Cancer Prevention. Adv Nutr. 2019; **10**(6):1012-1028. doi:10.1093/advances/nmz046

[55] Henning SM, Wang P, Carpenter CL, Heber D. Epigenetic effects of green tea polyphenols in cancer. Epigenomics. 2013;5(6):729-741. doi:10.2217/epi.13.57

[56] Stefanska B, Karlic H, Varga F, Fabianowska-Majewska K, Haslberger A. Epigenetic mechanisms in anti-cancer actions of bioactive food components--the implications in cancer prevention.

Br J Pharmacol. 2012;**167**(2):279-297. doi:10.1111/j.1476-5381.2012.02002.x

[57] Ko JH, Sethi G, Um JY, Shanmugam MK, Arfuso F, Kumar AP, Bishayee A, Ahn KS. The Role of Resveratrol in Cancer Therapy. Int J Mol Sci. 2017; **18**(12):2589. doi: 10.3390/ijms18122589. PMID: 29194365; PMCID: PMC5751192.

[58] Pratheeshkumar P, Sreekala C, Zhang Z, Budhraja A, Ding S, Son YO, Wang X, Hitron A, Hyun-Jung K, Wang L, Lee JC, Shi X. Cancer prevention with promising natural products: mechanisms of action and molecular targets. Anticancer Agents Med Chem. 2012; **12**(10):1159-84. doi: 10.2174/187152012803833035.

[59] Varoni EM, Lo Faro AF, Sharifi-Rad J, Iriti M. Anticancer Molecular Mechanisms of Resveratrol. Front Nutr. 2016; 3:8. doi:10.3389/fnut.2016.00008

[60] Ratovitski EA. Anticancer Natural Compounds as Epigenetic Modulators of Gene Expression. Curr Genomics. 2017; **18**(2):175-205. doi:10.2174/1389202917666160803165229

[61] Shukla S, Meeran SM, Katiyar SK. Epigenetic regulation by selected dietary phytochemicals in cancer chemoprevention. Cancer Lett. 2014; **355**(1):9-17. doi:10.1016/j.canlet.2014.09.017

[62] Sun Y, Jiang X, Chen S, Price BD. Inhibition of histone acetyltransferase activity by anacardic acid sensitizes tumor cells to ionizing radiation. FEBS Lett. 2006; **580**(18):4353-6. doi: 10.1016/j.febslet.2006.06.092. Epub 2006 Jul 10. PMID: 16844118.

[63] Samad N, Javed A. Therapeutic Effects of Gallic Acid: Current Scenario. J Phytochemistry Biochem 2018; 2: 113.

[64] Weng YP, Hung PF, Ku WY, Chang CY, Wu BH, Wu MH, Yao JY, Yang JR, Lee CH. The inhibitory activity of gallic acid against DNA methylation: application of gallic acid on epigenetic therapy of human cancers. Oncotarget. 2017; **9**(1):361-374. doi: 10.18632/oncotarget.23015.

[65] Kuo HD, Wu R, Li S, Yang AY, Kong AN. Anthocyanin Delphinidin Prevents Neoplastic Transformation of Mouse Skin JB6 P+ Cells: Epigenetic Re-activation of Nrf2-ARE Pathway. AAPS J. 2019; **21**(5):83. doi:10.1208/s12248-019-0355-5

[66] Ramasamy K, Agarwal R. Multitargeted therapy of cancer by silymarin. Cancer Lett. 2008; **269**(2): 352-362. doi:10.1016/j.canlet.2008.03.053

[67] Deep G, Agarwal R. Antimetastatic efficacy of silibinin: molecular mechanisms and therapeutic potential against cancer. Cancer Metastasis Rev. 2010; **29**(3):447-463. doi:10.1007/s10555-010-9237-0

[68] Cui W, Gu F, Hu KQ. Effects and mechanisms of silibinin on human hepatocellular carcinoma xenografts in nude mice. World J Gastroenterol. 2009; **15**(16):1943-1950. doi:10.3748/wjg.15.1943

[69] Kalaiselvan V, Kalaivani M, Vijayakumar A, Sureshkumar K, Venkateskumar K. Current knowledge and future direction of research on soy isoflavones as a therapeutic agents. Pharmacogn Rev. 2010;4(8):111-117. doi:10.4103/0973-7847.70900

[70] Kalaiselvan V, Kalaivani M, Vijayakumar A, Sureshkumar K, Venkateskumar K. Current knowledge and future direction of research on soy isoflavones as a therapeutic agents. Pharmacogn Rev. 2010; **4**(8):111-117. doi:10.4103/0973-7847.70900

[71] Pudenz M, Roth K, Gerhauser C. Impact of soy isoflavones on the epigenome in cancer prevention. Nutrients. 2014; **6**(10):4218-4272. doi:10.3390/nu6104218

[72] Spagnuolo C, Russo GL, Orhan IE, Habtemariam S, Daglia M, Sureda A, Nabavi SF, Devi KP, Loizzo MR, Tundis R, Nabavi SM. Genistein and cancer: current status, challenges, and future directions. Adv Nutr. 2015; **6**(4):408-19. doi: 10.3945/an.114.008052.

[73] Tuli HS, Tuorkey MJ, Thakral F, Sak K, Kumar M, Sharma AK, Sharma U, Jain A, Aggarwal V, Bishayee A. Molecular Mechanisms of Action of Genistein in Cancer: Recent Advances. Front Pharmacol. 2019; 10:1336. doi: 10.3389/fphar.2019.01336. PMID: 31866857; PMCID: PMC6910185.

[74] Verma M. Cancer control and prevention by nutrition and epigenetic approaches. Antioxid Redox Signal. 2012; **17**(2):355-364. doi:10.1089/ars.2011.4388

[75] Zhang, Y, Kutateladze TG. Diet and the epigenome. Nat Commun 2018; **9**, 3375

[76] Tiffon C. The Impact of Nutrition and Environmental Epigenetics on Human Health and Disease. Int J Mol Sci. 2018; 19(11):3425. doi:10.3390/ijms19113425

[77] Pehlivan, F. Vitamin C: An Epigenetic Regulator. In: Vitamin C: An Update on Current Uses and Functions. Jean Guy LeBlanc (Ed.) InTech Open. 2018. doi. 10.5772/INTECHOPEN.82563

Chapter 5

Epigenetic

Mehmet Ünal

Abstract

Lately, a brand-new studies agenda emphasizing interactions between societal elements and wellness has emerged. The phrase social determinant of health and fitness typically refers to any nonmedical element directly effecting health, including behaviors, knowledge, attitudes, and values. Status of health is adversely and strongly impacted throughout the life span by social disadvantages. Epigenetic mechanisms are implicated in the processes through which social stressors erode health in humans and other animals. Research in epigenetics suggests that alterations in DNA methylation might offer a temporary link between interpersonal adversity and wellness disparity. Likewise, accelerated loss in telomeres is extremely correlated not only with chronic and social stress but also aging. Therefore, it may provide a link between the various physiological events associated with health inequalities. Research in epigenetics indicates that alterations in DNA methylation may provide a causal link between social adversity and health disparity. Additionally, these experimental paradigms have yielded insights into the potential role of epigenetic mechanisms in mediating the effects of the environment on human development and indicate that consideration of the sensitivity of laboratory animals to environmental cues may be an important factor in predicting long-term health and welfare.

Keywords: stress, epigenetic, DNA, social, environment

1. Introduction

Absence of early life stress has great effects on both health and well-being. This particular topic focuses on the availability or quality of food sources, exposure to toxic effects, community-based events, and the presence of stressors or threats in the ecosystem. In humans, years of extensive researches have revealed some sort of correlation among variations within being exposed in early life era and long-lasting risks of psychiatric and physical illness of maturity [1].

While it is expected that the effects of exposure to chemical substances besides deprivation or stress hormones of a vital vitamin or energy supply, they might have biological impacts on human depending on the specification of the exposure and availability. Personal and social experiences carry the partly same features with biochemical, cellular, and neurobiological changes and this case is amplified with psychosocial expertise executed on animal models in the light of finding answers to these questions.

Effects of social experiences and aggravating events taken place in life leads to the experimental study which has mainly focused on types of lab rodents (usually including mice and rats), albeit several numbers of primate studies can be obtained to maintain additional data in the interest of powerful impacts of encounters taken place in the initial stages of life [2, 3]. In these designs, impact of prenatal nervousness, absence of a mother role and also disengagement variance in maternal treatment, adolescent cultural enhancement, isolation and mature cultural anxiety. When these have been investigated, results indicate that the quality of emotional stress experience or social experiences may cause neuroendocrine consequences which effects social and reproductive behavior. While it is the evidence of there is a phase of increased sensitivity to these eco-induced factors during prenatal as well as first postnatal development, there may also be plasticity that extends into adulthood and adolescence after infancy. The long-term negative impact of the first latter life events on human brain's region-specific gene expression part would be an important finding within these studies. Substantial evidence supports the existence of longterm effects on HPA axis functioning following early life stress; these effects persist into adulthood and are accompanied by lasting behavioral changes. In clinical studies, early life stress has been shown to be a strong predictor of ACTH responsiveness [4, 5]. Further analysis of the molecular mechanisms that could mediate this long-lasting effect involved in gene regulation have been led by these findings.

Epigenetic processes are effective events in changing gene expression and the long-term effects of events experienced in the early stages of life on gene expression are the subject of biologically based research [6]. Across species, it is obvious that a selection of experiences, such as the aspects of interactions experienced socially as well as being exposed to the stressors, can trigger epigenetic consequences. In addition, these progressive outcomes may be carried over descendants in certain cases, resulting in behavioral and neurobiological disturbance of the offspring and even of the grand offspring [7, 8].

2. Epigenetic effects of stress experienced in early life

The crucial role of epigenetic machinery in the biological embedding of stressful exposures in early life has been demonstrated in a number of rodent models, where considerable variations in both DNA methylation and histone modification have been reported in offspring exposed to different prenatal stresses, inappropriate maternal care, maternal deprivation/separation, as well as to juvenile social enrichment/isolation [9–11]. It was found out by the rodent tests that mothers' offspring with comparatively flat postnatal maternal hygiene grades increased fearfulness, anxiety, and stress-reactivity compared to the mothers' offspring with caregiving of normal PH grades [12]. These behavioral anomalies were followed by diminished exposition of the glucocorticoid receptor (GR) encoding NR3C1 gene within the hippocampus along with increased methylation of CpG dinucleotides in the NR3C1 promoter and additionally great levels of activation of the hypothalamic–pituitary–adrenal (HPA) axis and serum glucocorticoids [13–15]. Maternally deprived offspring developed irritability, anxiety, depressive symptoms, interruption of socially received interactions, genome-wide changes and high reactivity of DNA methylation transcription [16–18].

Depriving the organization and maternal attention of some other family members of juvenile rhesus macaques and infants also induced stress and depression related symptoms followed by protracted activation of the HPA axis. Consecutively, changed genome-wide modifications and gene transcription to DNA methylation patterns are within the mental faculties and in the peripheral T lymphocyte [19–24].

The amygdala, hippocampus, HPA axis, and medial prefrontal cortex form the areas affected by reduced maternal care in rodents [25]. Within the very first week of post-natal existence, maltreatment of mothers breastfeeding new-born children has shown to cause frequent vocalized distress calls in offspring and shifts in patterns of methylcytosine last long and its hydroxymethyl cytosine derivative at many spots such as the BDNF locus, in the amygdala, hippocampus, and in exposed offspring's medial prefrontal cortex, all at the time of exposure [26–28]. Early-life stress as in maternal separation-induced in the paraventricular nucleus of the hypothalamus increased AVP and POMPC expression and decreased DNA methylation at these loci [29]. Experienced early-life stress increased NR3CR1 transcription in the paraventricular nucleus and rendered CRH transcription refractory to maturity chronical stress after stress experienced in early life, suggesting that it jeopardizes CRH transcriptional responsiveness in the hypothalamus to later chronic stressors, probably through a GR-dependent mechanism [30].

The long-lasting effects of prenatal maternal pressure on the neural genes as well as offspring behavioral word was showed more by a research analyzing the effects on adult offspring of prenatal exposure to predator odor during fetal gestation. In offspring of whose mothers suffered being exposed to the odor of predator in the time of pregnancy and in female (yet no male) offspring, stated endocrine and behavioral modifications were followed by BDNF's diminished expression and improved DNA methylation of BDNF promoter sequences of the hippocampus, as well as bb promoter sequences of the hippocampus, odor avoidance, addition to corticosterone production as a response to being exposed to the odor, were enhanced. Mentioned studies show that not only prenatal but also postnatal stressors may result in long-lasting alterations in epigenetic control, neuroendocrine function, and neural gene transcription activity that continuing through adulthood [31].

Research one on humans have advanced and expanded the results of animal studies on the effects of early-life strains [32]. So, proof states that adolescence, infancy, and fetal gestation are actually delicate stages during which epigenetic, psychological, and behavioral changes continue through adulthood which may be caused by exposing to cultural adversity. A recent systematic review reported, forty studies which were betwixt 2004 and 2014, that listed NR3C1 methylation changes in response to early life adversity, parental anxiety, and psychopathology studies, of which twenty-seven were human studies [33]. While in these papers a variety of different NR3C1 sequences are actually involved as reglementary targets, perhaps the most stable finding is actually a highly related expansion in exon 1F methylation in the NR3C1 gene of the individual (or maybe the analogous exon seventeen in the thirteen animal studies) as well as early life adversity experience. Exon 1F/17 has a portion of the DNA sequence encoding a methylation-sensitive binding website for the NGF1A/ EGR1 [34] controlled transcription activator for neural activity. The decreased NR3C1 expression caused by increased methylation of this unique binding website, consecutively decreases the means of providing bad feedback to the hypothalamus and pituitary mediated by glucocorticoid, resulting in continuous activation of the HPA axis, as well as the ensuing disorders. Improved NR3C1 promoter methylation has been associated with a variety of experiences highly appropriate for inducing prenatal tension, such as near partner aggression, or perhaps maternal exposure to genocidal war [35, 36]. In addition, neglect, being abused in childhood, and destitute were also concluded to be related with enhanced methylation of the NR3C1 promoter [34, 37–39]. Recently prepared reports give further backing for a correlation among early life adversity, improved methylation of the promoter, and reduced NR3C1 transcription [38, 40–44]. These mixed scientific studies indicate that as a conse-quence of allostatic overload caused by a range of stressful interactions, attenuation of the NR3C1 phrase is actually an aspect of the task leading to elevation of HPA axis operation. Despite such insights, it remains unclear how *NR3C1* is specifically tar-geted for epigenetic silencing, and whether methylation of the NGF1A/EGR1 binding site, which prevents NGF1A/EGR1 recruitment to *NR3C1*, is accompanied by other changes that attenuate *NR3C1* expression in the brain. As inhibitors of NR3C1 protein run [45, 46], noncoding RNAs such as the lncRNA GAS5 have been involved, as well as miRNAs such as MIR 124 could as well control the durability of NR3C1 transcripts [47, 48]. Though, several different early life adversity response studies including DNA methylation changes linked to several extra-human genes, such as SLC17A3, PM20D1, KITLG, SLC6A4, BDNF, MORC1, LGI1/LGI2, FKBP5, CRHBP, CRH, and MAOA [49–54]. Additional studies to explain the purposeful interrelationships of genes with NR3C1, along with a greater comprehension of exactly how these genes are actually controlled, possibly make it possible to reveal precisely how prenatal stress, child-hood neglect, and parental hygiene deprivation are biologically embedded and have long-lasting effects that dwell on across. In addition, an in-depth study of the proce-dures may explicate the biological base of resistance to anxiety as well as provide an evidence base for successful interventions.

3. Epigenetic plasticity in adulthood as well as adolescence

While plasticity carrying epigenetic pathways was originally considered to be confined for the premature steps of embryogenesis, now it is becoming more and more apparent to that epigenetic variation can be caused by experiences occurring during the lifespan. In addition, a key facet of memory and learning from infancy to adulthood may be the guiding source to modify DNA methylation and histone tails [55]. Variation of epigenetic, similarly, has been related with the alterations in phenotype and gene expression in the mean time of the latter stages of advancement in the form of studies on the effect of social experiences & stressors. In addition to adult mice with genetically mediated memory dysfunction, being exposed to complicated housing environments for four weeks were found to be associated with enhanced hippocampal histone acetyla-tion as well as memory enhancement and cortex [56]. Surprisingly, these enrichment-induced effects on histones, memory and comprehending, can as well be obtained with pharmacological remedies in non-memory-impaired mice that promote histone acetylation. Histone modifications (particularly histone methylation) are also observed in mice in enriched settings in the BDNF III, IV, and VI promoter regions [57].

Chronic stress is actually linked to decreases in the expression of this specific gene compared to environmental enrichment, which has been shown to advance stages of BDNF, an epigenetic base might be an issue for such consequences. In rats, immobi-lization pressure was found to cause great rises in hippocampal DNA methylation in the BDNF gene, coupled with exposure to predator odor [58]. Reductions in BDNF are actually found within the version of social defeat after a month of being exposed to what is known as social stressor, then hippocampal histone demethylation at the promoter of BDNF might take credit for such specific impact [59]. Histone acetyla-tion is temporarily diminished, after that shows elongated risings on mice that is rejected socially. Therefore, such specific impact may be related with long-lasting reductions in histone deacetylase amounts induced by stress [60]. Similarly, increased histone acetylation in rats continues to be observed for up to twenty-four hours following the experience of repeated population-based rejection [61]. In addition, the behavior-based outcomes of interpersonal loss, such as diminished social-based activity, can be pharmacologically turned around thanks to a medication that inhibits histone deacetylases [60]. It is concluded that a group of mice were resistance to our stressor. And the other conclusion is that discomfiture between humans results in consequences which would last long. Stress-susceptible mice were found in a recent study to have elevated levels of CRF mRNA in the PVN and also diminished methylation of DNA in the CRF gene [62]. In comparison, stress-resilient mice were found to undergo no changes in the CRF mRNA or perhaps DNA methylation of this particular gene. In fact, differential sensitivity to the effects of stress is a key thing to consider in these studies, and as a consequence of anxiety, there is growing evidence for strain or perhaps genotype-specific epigenetic consequences [63].

4. Transgenerational effect of the social stress and environment

The continuous epigenetic effects of environmental events were presented in the previous sections. In influencing behavioral and neurobiological effects, these consequences seem to play a crucial role. However, it may also be the case that these eco-induced influences are capable of persisting over centuries. One mechanism by which this transmission occurs [64] may be transgenerational continuity in maternal behavior. There is evidence in rodents that coercive caregiving as well as variance in maternal LG can change the enhancement of female offspring so that each offspring as well as grand offspring can also find these maternal characteristics. In the medial preoptic position of the hypothalamus of female offspring, maternal LG tends to alter the DNA methylation as well as the Esr1 phrase [65]. Throughout the postnatal period, during which adulthood continues, these implications emerge. ER-alpha quantities possess a crucial part in deciding the responsiveness of females to circulating oestrogens for late gestation, and stated specific responsiveness predicts the quality as well as the number of interactions between postnatal mother babies. Individual differences in maternal LG are actually transferred from mother to offspring (generation F1) as well as to grand offspring (generation F2) as a result of the epigenetic changes [64, 66].

In studies of child violence, variance of behavioral segmentations is found across species. In rats, both the variation of DNA methylation and the BDNF phrase in the prefrontal cortex can be correlated with the transgenerational continuity of violence. Females' BDNF term tends to be decreased if experienced violence in childhood and methylation of BDNF IV promoter DNA in the prefrontal cortex is to be improved by this situation as well. The offspring of these females likewise have increased Bdnf IV promoter DNA methylation in the prefrontal cortex. Addition to these, offspring of stated females is likely to have enhanced methylation of BDNF IV promoter DNA in the prefrontal cortex [67]. Surprisingly, scientific studies on cross-fostering recommend that prenatal elements may be linked to the transmission of violence and also the epigenetic variance related to this specific phenotype rather than postnatal events with the dam. In fact, the direct inheritance of epigenetic change is another path by which perinatal epigenetic, as well as behavioral effects, may persist over generations. Although it has long been believed that during the first stages of embryogenesis, a complete erasure of epigenetic disruption to the genome is at hand. Revelation of imprinted genes (genes that are expressed depending on the parent from whom they are inherited) has led to widely believed speculation that the previous generation's epigenetic "memory" is actually retained and transmitted to the genome [68].

The epigenetic inheritance of the toxicological exposure effect has been studied and also suggests the existence of the patriline effects for many decades [69]. More recently, in offspring exposed to separation (F1), the offspring of separated males (F2), and the grand offspring of separated males (F3), the transgenerational impact of maternal separation have been investigated. With this paradigm, during the postnatal era, mice were exposed to unforeseen separation from the dam. Depressive-like behaviors that were activated in the offspring of the F1 model were found to be contained in both F2 and F3 generation mice [70]. DNA methylation patterns caused by separation have been shown to be contained in the sperm and brains of F1 male mice and also in the brains of F2 males. Hypermethylation of the Mecp2 gene as well as hypomethylation of the Crfr2 gene resulted in these epigenetic consequences. As a result, it appears that the epigenetic variation that is created with the early life environment's aspect could be encoded within the germ cells, leaving traces of stress and social environment on the next descendants through transgenerational approach.

5. Human health's social-based determinants

A distinguishing figure in inequal cultures is the fact that in neighborhoods and neighborhoods with lower socioeconomic mobility (SES) [71, 72], community problems and persistent wellbeing appear to be more prevalent. In lower SES populations, the risks of illnesses such as cardiovascular disease, stroke, diabetes, obesity, and mental conditions are highest. It as well needs to be remembered, however, that the prevalence of such chronic, non-communicable diseases is actually classified across social classes, with the lowest incidence in higher SES categories. Expectancy of life across social classes is equally ranked, being higher in high SES categories. The prospect of designing evidence-based health and policy that covers social subjects strategies is to elucidate the biological and social roots of the social gradient in health that could increase well-being and health for everyone while providing additional benefit to those with higher needs. Good knowledge of the biochemical mechanisms by which wellbeing is impacted by the social gradient will also help to identify relevant intervention or mitigation goals and sustain biomarkers for which to track effects.

Salary inequality can be seen as sensitive, quantitative relative place measures among a larger hierarchy of socioeconomic status that indicate disparities in access to, along with economic resources, a number of forms of social, educational and cultural capital [73–75]. It has been speculated that social rank is essentially a result of the amount of entrance obtained to stated and differentiated forms of resources in extremely unequal environments that competing social experiences for such access are actually mental stressors that can contribute to uncertainty about ranking. Many scientific researches confirm the theory of status anxiety, connecting not so high social expectations to nervousness, guilt, depression and harming oneself [76–79]. In addition, status-based nervousness and its not only mental but also cognitive outcomes are potentially possible contributors to some kinds of social adversity, childhood era deprivation for example, limited autonomy on making decisions which even consider matter of life events, reduced social connectivity, and decreased levels of interest in certain different parts and members of the community. Such theories align with the findings of hierarchies of animal domination, yet in which situation that dominant individuals are literally prohibited from ensuring preferential entrance to group services provided in scarce supplies, such as meals, water, accommodation, as well as companions, and access to subordinates. As previously stated, the hierarchy of dominance is currently an evolutionarily and strongly maintained form of social organization. Undoubtedly, social rank understanding evolves very quickly in members of humanity, as it is a prevalent conception for infants [hundred] and is used by children from two years of age to create relationships of dominance [80, 81].

As potential causes of chronic behavioral tension and allostatic load, inducing hypercortisolaemia and improved levels of inflammatory biomarkers in the blood, the pervasive experiences of status tournaments, in which rank is actually guided and controlled by oneself and others, have been labeled [82, 83]. Moreover, epidemiological data suggest that low SES increases allostatic load and increases blood inflammatory biomarkers [84, 85]. Therefore, the tendencies of too many social challenges to erode group harmony, degrade social networking sites and impede mutual assistance, such as persistently competitive activities and violence, may place restrictions on social practices that decide the fundamental characteristics of human well-being.

Emerging research recognizes social networking platforms as critical factors in health security [86–90] and supports the theory that loss of social capital accentuates the very poor health outcomes of low SES communities by weakening or even diminishing social networks. In order to minimize allostatic burden and buffer the immune system against inflammatory stimuli associated with very low SES [91–94], increased parental assistance has been confirmed. Close comparisons can be drawn between the mechanisms of action of animal and human social buffering treatments, general themes of which include reducing the function of the HPA axis, attenuating inflammation and increasing the development of oxytocin [95]. Via enhancing parenting skills, strengthening family relationships, or even developing capacities for young people, social interventions that promote group buffering have all been found to decrease pro-inflammatory biomarkers, indicating a preventive impact, while some aspects of resilience-building may be much more durable relative to others. Additional analyses of these and other human cohorts would enable the biological mechanisms and psychosocial processes by which such strategies accomplish their buffering effects to be explained in greater detail [93, 94].

6. The epigenetic effects of persistent interpersonal stress in humane communities

In particular being long-lasting exposed to the social-based stressors associated with low SES over the course of life are recognized to impact the likelihood of chronical illness by using their effects on a broad variety of physiological processes affecting the nervous, hemopoietic, cardiovascular, and endocrine systems. Most documented studies of the effects of SES on the epigenome showed that DNA methylation patterns in samples taken from tissues collected, such as whole blood, fractionated white blood cells, or perhaps buccal swabs were altered. Numerous scientific works indicate that SES is actually connected to differences in patterns of genomic DNA methylation [96, 97]. Childhood SES was shown to be specifically correlated with differential methylation of 1252 gene promoters in forty individuals from the 1957 British Birth Cohort in one of several experiments using promoter microarrays so that increased methylation was related with lower childhood SES for 586 promoters and superior childhood SES for the remaining 666 promoters [96]. A review of 239 participants of the Glasgow pSoBid cohort, which indicates a particularly steep social gradient in wellbeing, found with an alternative technique of DNA methylation analysis that average DNA methylation levels across the entire genome were more or less seventeen percent lower in the most deprived group than in perhaps the least deprived group [98]. It remains to be clarified if this low SES-related hypomethylation of whole genomic DNA is specifically directed at gene bodies, cis-regulatory elements, inter-genic areas, and/or repetitive DNA sequences. Recent studies have reported that genes whose transcription is actually involved in neuroendocrine and inflammatory responses to low SES also show epigenetic SES sensitive modifications, which are actually in line with previous studies involving process dysregulation in low SES individuals [99, 100].

In more unequal environments, possibility of mental wellness problems such as nervousness, alcohol, and anxiety are actually much higher. Consecutively, these disorders are certainly easier to come across in low-SES communities [101]. Predictive connections between lower SES, differential methylation of the SLC6A4 serotonin transporter gene promoter, greater amygdala activity, as well as signs of depression have been reported in a recently published analysis [102]. Previously, differential SLC6A4 methylation has been shown to be exclusively linked to molestation of child [103], low SES [104], stress-related depression [105], as well as enhanced amygdala reactivity to frightening stimuli. In addition, increased fear reactivity of the amygdala in puberty has been shown to become a likely biomarker indicative of adult stress and tension [106–108]. These findings were expanded to include a possible analysis of depression development within a cohort of teenagers tested on 3 occasions at 11–15,13–18 and 14–19 years of age, in order [102]. Using blood and saliva samples for DNA methylation analysis, low SES at age 11–15 years was discovered to be predic-tive of increased methylation of SLC6A4 at age 13–18 years, which was predictive of improved amygdala reactivity to a fearful stimulus (detected by fMRI) over exactly the same time, and which subsequently was connected with an elevated threat of depression between ages of 14–19 for adolescents with an optimistic depression-based family history. Such outcomes, therefore, propose a plausible biological path through which low SES, by methylation of SLC6A4, may attenuate expression of the serotonin transporter encoded by that stated particular gene, forwarding to elevations not only in amygdala reactivity but also in liability of depression. Additionally, such indica-tions pinpoint possible biomarkers for building and evaluating preventive or maybe treatment-based interventions that may buffer the effects of low SES on liability of depression in the means of lifetime.

7. Post-traumatic stress and anxiety

The intense stress caused by imposed to stressful experiences, battle for example, genocide, starvation, elevates the liability of psychiatric wellbeing problems, includ-ing PTSD, schizophrenia, depression, even killing oneself [36, 37]. Similarly, US fighting veterans with an analysis of moderate PTSD showed hypomethylation on the regulatory portion of the NR3C1 promoter 1F and decreased HPA axis actions [109]. As a result, in the meantime NR3C1 is a very typical target for epigenetic modifica-tions as a response upon these distinctive stress forms, in addition to qualitative dis-crepancies in the mechanisms as well as the context of traumatic events themselves, the particular variations within the patterns of NR3C1 methylation could represent disparities in the timings as well as times of coverage for trauma.

Traumatic stress and PTSD-associated variance in DNA methylation at the SKA2 locus have been reported in recently made studies on epigenetic alterations among war vets suffering extreme PTSD [110, 111]. SKA2 encodes a protein that is rather to serve as a chaperone or possibly a GR activity regulator, allowing negative feedback to the HPA axis mediated by cortisol dependent GR. SKA2 was recognized as a hyper-methylated, under expressed locus of suicide completers in post-mortem cortical tissue, and variance in SKA methylation was also correlated with suicidal activities in PTSD individuals [112, 113]. Although the discovery of altered DNA methylation taken from those who tried to commit suicide and who suffers PTSD at this locus in tissue indicates possible functions for SKA2 in the control of traumatic stress response, it is presently uncertain if these modifications are directly linked to the multiple psychopathological behaviors under investigation.

In intergenerational epigenetic responses to stress, research on the effects of PTSD suffered by those who survived Holocaust through their offspring includes modified HPA axis actions, and even more especially dysregulation of the GR along with its auxiliary components. In a similar study, methylation with a CpG dinucleotide inside a GR binding website inserted within an intron of this gene encoding the GR regula-tor FKBP5 was shown to be increased within the blood cells of Holocaust survivors, as well as lower in their offspring, relative to the quantity of this gene encoding the GR regulator FKBP5 [114]. These findings show that intergenerational transfer of trauma-related DNA methylation modifications has taken place between Holocaust survivors and descendants of theirs in at least a variety of cases. In response to a new, moderately demanding exposure, diminished behavioral regulation was recognized as diminished latency to type in unknown places, indicative of probably elevated resistance or impulsivity. Additionally, DNA methylation switches at candidate gene loci transmitted via the germline along with mediated, transmitted behavioral anomalies from the MSUS. The applicant genes include Nr3c1, which demonstrated decreased methylation of the promoter and increased transcription of this progeny of MSUS treated mice in the hippocampus. In addition, traumatized men's sperm contained microinjection and trauma-induced miRNAs of distilled RNA derived from their traumatized adult men's sperm recapitulated MSUS-induced behavioral abnormalities, suggesting functions within the intergenerational transmission of trauma-induced phenotypes for these very short noncoding RNAs. Interestingly, the intergenerational transfer of MSUS mediated altered behavioral reactions to moder-ately challenging stimuli, increased Nr3c1 methylation, and decreased transcription of this specific gene within the hippocampus was improved by environmental enrich-ment. In a related study, corticosterone administration to adult male mice caused behavioral phenotypes that indicated hyper nervousness within their male F1 prog-eny, as well as reduced anxiety levels, but elevated depressed characteristics within their F2 progeny. In addition, both the F1 and F2 phenotypes have been identified with the expression within the paternal sperm of some corticosterone mediated miRNAs. Taken together, these animal experiments suggest that ameliorative and negative interactions modulate behavior, the effects of which could be delivered among one descendant to the others, possibly by epigenetic modulation of neuroen-docrine reaction systems involving miRNAs [115].

8. Conclusion

The pervasive influence of social stressors on well-being and health are recorded in detailed literature. Emerging data shows that the biological embedding of psy-chosocial messages throughout the body potentially includes epigenetic pathways, influencing cellular mechanisms that could influence health hazards over the life cycle. Many vertebrates display social activities that represent similar operating concepts of hierarchical domination, like teleost fish, non-human primates, and humans, and elicit stress responses that are actually followed by impacts on the epigenome. In hierarchies, social differences can act as persistent behavioral stressors and limit access to services, leading to a social gradient in health and expectancy of life in humans. Epigenetic changes are triggered by a broad scope of behavior-based stressors, traumatically suffered experiences included which present promise as biomarkers of disease risk and can likely be reversed by intervference to lessen the health-based effects of distress.

Conflict of interest

The authors state no conflict of interest.

Author details

Mehmet Ünal
Psychoneuroimmunology, Kocaeli, Turkey

*Address all correspondence to: fztmehmet6@gmail.com

IntechOpen

References

[1] Tomalski P, Johnson MH. The effects of early adversity on the adult and developing brain. Curr Opin Psychiatry. 2010;23(3):233-238.

[2] Harlow HF, Dodsworth RO, Harlow MK. "Total social isolation in monkeys." Proceedings of the National Academy of Sciences of the United States of America 54.1 (1965): 90.

[3] Suomi SJ, et al. Effects of maternal and peer separations on young monkeys. J Child Psychol Psychiatry. 1976;17(2): 101-112.

[4] Ladd CO, et al. Long-term adaptations in glucocorticoid receptor and mineralocorticoid receptor mRNA and negative feedback on the hypothalamo-pituitary-adrenal axis following neonatal maternal separation. Biol Psychiatry. 2004;55(4):367-375.

[5] Maccari, Stefania, et al. "Prenatal stress and long-term consequences: implications of glucocorticoid hormones." Neuroscience & Biobehavioral Reviews 27.1-2 (2003): 119-127.

[6] Curley JP, et al. Social influences on neurobiology and behavior: epigenetic effects during development. Psychoneuroendocrinology. 2011;36(3): 352-371.

[7] Champagne FA. Epigenetic mechanisms and the transgenerational effects of maternal care. Front Neuroendocrinol. 2008;29(3): 386-397.

[8] Franklin TB, et al. Epigenetic transmission of the impact of early stress across generations. Biol Psychiatry. 2010;68(5):408-415.

[9] Anacker, Christoph, Kieran J. O'Donnell, and Michael J. Meaney. "Early life adversity and the epigenetic programming of hypothalamic-pituitary-adrenal function." Dialogues in clinical neuroscience 16.3 (2014): 321.

[10] Maccari S, et al. The consequences of early-life adversity: neurobiological, behavioural and epigenetic adaptations. J Neuroendocrinol. 2014;26(10):707-723.

[11] Kundakovic M, Champagne FA. Early-life experience, epigenetics, and the developing brain. Neuropsychopharmacology. 2015;40(1):141-153.

[12] Weaver IC, Cervoni N, Champagne FA, et al. Epigenetic programming by maternal behavior. Nat Neurosci. 2004;7(8):847-854.

[13] Weaver IC, Meaney MJ, Szyf M. Maternal care effects on the hippocampal transcriptome and anxiety-mediated behaviors in the offspring that are reversible in adulthood. Proc Natl Acad Sci USA. 2006;103(9):3480-3485.

[14] McGowan PO, Suderman M, Sasaki A, et al. Broad epigenetic signature of maternal care in the brain of adult rats. PLoS One. 2011;6(2):e14739.

[15] Suderman M, McGowan PO, Sasaki A, et al. Conserved epigenetic sensitivity to early life experience in the rat and human hippocampus. Proc Natl Acad Sci USA. 2012;109 Suppl. 2:17266-17272.

[16] Murgatroyd C, Patchev AV, Wu Y, et al. Dynamic DNA methylation programs persistent adverse effects of early-life stress. Nat Neurosci. 2009;(12): 1559-1566.

[17] Franklin TB, Russig H, Weiss IC, et al. Epigenetic transmission of the impact of early stress across generations. Biol Psychiatry. 2010;68(5):408-415.

[18] Franklin TB, Saab BJ, Mansuy IM. Neural mechanisms of stress resilience and vulnerability. Neuron. 2012;75(5): 747-761.

[19] Provencal N, Suderman MJ, Guillemin C, et al. The signature of maternal rearing in the methylome in rhesus macaque prefrontal cortex and T cells. J Neurosci. 2012;32(44): 15626-15642.

[20] Cole SW, Conti G, Arevalo JM, Ruggiero AM, Heckman JJ, Suomi SJ. Transcriptional modulation of the developing immune system by early life social adversity. Proc Natl Acad Sci USA. 2012;109(50):20578-20583.

[21] Suomi SJ. Risk, resilience, and gene-environment interplay in primates. J Can Acad Child Adolesc Psychiatry. 2011;20(4):289-297.

[22] Dettmer AM, Suomi SJ. Nonhuman primate models of neuropsychiatric disorders: influences of early rearing, genetics, and epigenetics. ILAR J. 2014;55(2):361-370.

[23] Kinnally EL. Epigenetic plasticity following early stress predicts long-term health outcomes in rhesus macaques. Am J Phys Anthropol. 2014;155(2):192-199.

[24] Nieratschker V, Massart R, Gilles M, et al. MORC1 exhibits cross-species differential methylation in association with early life stress as well as genome-wide association with MDD. Transl Psychiatry. 2014;4:e429.

[25] McEwen BS, Nasca C, Gray JD. Stress effects on neuronal structure: hippocampus, amygdala, and prefrontal cortex. Neuropsychopharmacology. 2016;41(1):3-23.

[26] Blaze J, Scheuing L, Roth TL. Differential methylation of genes in the medial prefrontal cortex of developing and adult rats following exposure to maltreatment or nurturing care during infancy. Dev Neurosci. 2013;35(4): 306-316.

[27] Roth TL, Matt S, Chen K, Blaze J. Bdnf DNA methylation modifications in the hippocampus and amygdala of male and female rats exposed to different caregiving environments outside the homecage. Dev Psychobiol. 2014;56(8): 1755-1763.

[28] Doherty TS, Forster A, Roth TL. Global and gene-specific DNA methylation alterations in the adolescent amygdala and hippocampus in an animal model of caregiver maltreatment. Behav. Brain Res. 298(Pt A), 55-61 (2016).

[29] Wu Y, Patchev AV, Daniel G, Almeida OF, Spengler D. Early-life stress reduces DNA methylation of the Pomc gene in male mice. Endocrinology. 2014;155(5):1751-1762.

[30] Bockmuhl Y, Patchev AV, Madejska A, et al. Methylation at the CpG island shore region upregulates Nr3c1 promoter activity after early-life stress. Epigenetics. 2015;10(3):247-257.

[31] St-Cyr S, McGowan PO. Programming of stress-related behavior and epigenetic neural gene regulation in mice offspring through maternal exposure to predator odor. Front. Behav. Neurosci. 9, 145 (2015). 27 Andersen SL. Exposure to early adversity: points of crossspecies translation that can lead to improved understanding of depression. Dev. Psychopathol.(2), 477-491 (2015).

[32] Turecki G, Meaney MJ. Effects of the social environment and stress on

glucocorticoid receptor gene methylation: a systematic review. Biol Psychiatry. 2016;79(2):87-96.

[33] A comprehensive review of the literature describing experience-dependent changes in DNA methylation patterns at the NR3C1 locus in humans and model organisms.

[34] McGowan PO, Sasaki A, D'Alessio AC, et al. Epigenetic regulation of the glucocorticoid receptor in human brain associates with childhood abuse. Nat Neurosci. 2009;12(3):342-348.

[35] Radtke KM, Ruf M, Gunter HM, et al. Transgenerational impact of intimate partner violence on methylation in the promoter of the glucocorticoid receptor. Transl Psychiatry. 2011;1:e21.

[36] Mulligan CJ, D'Errico NC, Stees J, Hughes DA. Methylation changes at NR3C1 in newborns associate with maternal prenatal stress exposure and newborn birth weight. Epigenetics. 2012;7(8):853-857.

[37] Labonte B, Yerko V, Gross J, et al. Differential glucocorticoid receptor exon 1(B), 1(C), and 1(H) expression and methylation in suicide completers with a history of childhood abuse. Biol Psychiatry. 2012;72(1):41-48.

[38] Tyrka AR, Parade SH, Eslinger NM, et al. Methylation of exons 1D, 1F, and 1H of the glucocorticoid receptor gene promoter and exposure to adversity in preschool-aged children. Dev Psychopathol. 2015;27(2):577-585.

[39] Weder N, Zhang H, Jensen K, et al. Child abuse, depression, and methylation in genes involved with stress, neural plasticity, and brain circuitry. J Am Acad Child Adolesc Psychiatry. 2014;53(4): 417-424.e415.

[40] van der Knaap LJ, Riese H, Hudziak JJ, et al. Glucocorticoid receptor gene (NR3C1) methylation following stressful events between birth and adolescence. The TRAILS study. Transl Psychiatry. 2014;4:e381.

[41] Braithwaite EC, Kundakovic M, Ramchandani PG, Murphy SE, Champagne FA. Maternal prenatal depressive symptoms predict infant NR3C1 1F and BDNF IV DNA methylation. Epigenetics. 2015;10(5):408-417.

[42] Murgatroyd C, Quinn JP, Sharp HM, Pickles A, Hill J. Effects of prenatal and postnatal depression, and maternal stroking, at the glucocorticoid receptor gene. Transl Psychiatry. 2015;5:e560.

[43] Radtke KM, Schauer M, Gunter HM, et al. Epigenetic modifications of the glucocorticoid receptor gene are associated with the vulnerability to psychopathology in childhood maltreatment. Transl Psychiatry. 2015;5:e571.

[44] Kertes DA, Kamin HS, Hughes DA, Rodney NC, Bhatt S, Mulligan CJ. Prenatal maternal stress predicts methylation of genes regulating the hypothalamic–pituitary–adrenocortical system in mothers and newborns in the Democratic Republic of Congo. Child Dev. 2016;87(1):61-72.

[45] Kino T, Hurt DE, Ichijo T, Nader N, Chrousos GP. Noncoding RNA gas5 is a growth arrest- and starvationassociated repressor of the glucocorticoid receptor. Sci Signal. 2010;3(107):ra8.

[46] Hudson WH, Pickard MR, de Vera IM, et al. Conserved sequence-specific lincRNA-steroid receptor interactions drive transcriptional repression and direct cell fate. Nat Commun. 2014;5:5395.

[47] Vreugdenhil E, Verissimo CS, Mariman R, et al. MicroRNA 18 and 124a down-regulate the glucocorticoid receptor: implications for glucocorticoid responsiveness in the brain. Endocrinology. 2009;150(5):2220-2228.

[48] Pan-Vazquez A, Rye N, Ameri M, et al. Impact of voluntary exercise and housing conditions on hippocampal glucocorticoid receptor, miR-124 and anxiety. Mol Brain. 2015;8:40.

[49] Melas PA, Wei Y, Wong CC, et al. Genetic and epigenetic associations of MAOA and NR3C1 with depression and childhood adversities. Int J Neuropsychopharmacol. 2013;16(7): 1513-1528.

[50] Cao-Lei L, Massart R, Suderman MJ, et al. DNA methylation signatures triggered by prenatal maternal stress exposure to a natural disaster: Project Ice Storm. PLoS One. 2014;9(9):e107653.

[51] Khulan B, Manning JR, Dunbar DR, et al. Epigenomic profiling of men exposed to early-life stress reveals DNA methylation differences in association with current mental state. Transl Psychiatry. 2014;4:e448.

[52] Non AL, Hollister BM, Humphreys KL, et al. DNA methylation at stress-related genes is associated with exposure to early life institutionalization. Am J Phys Anthropol. 2016;161(1):84-93.

[53] Houtepen LC, Vinkers CH, Carrillo-Roa T, et al. Genomewide DNA methylation levels and altered cortisol stress reactivity following childhood trauma in humans [•• Evidence is presented which indicates that increased DNA methylation at the KITLG locus is a mediator of childhood trauma-induced blunting of the cortisol response to stress in adulthood.]. Nat Commun. 2016;7:10967.

[54] Suderman M, Borghol N, Pappas JJ, et al. Childhood abuse is associated with methylation of multiple loci in adult DNA. BMC Med Genomics. 2014;7:13.

[55] Miller CA, Sweatt JD. Covalent modification of DNA regulates memory formation. Neuron. 2007;53:857-869.

[56] Fischer A, Sananbenesi F, Wang X, Dobbin M, Tsai LH. Recovery of learning and memory is associated with chromatin remodelling. Nature. 2007;447:178-182.

[57] Kuzumaki N, Ikegami D, Tamura R, Hareyama N, Imai S, Narita M, et al. Hippocampal epigenetic modification at the brainderived neurotrophic factor gene induced by an enriched environment. Hippocampus. 2011;21:127-132.

[58] Roth TL, Zoladz PR, Sweatt JD, Diamond DM. Epigenetic modification of hippocampal Bdnf DNA in adult rats in an animal model of posttraumatic stress disorder. J Psychiatr Res. 2011;45:919-926.

[59] sankova NM, Berton O, Renthal W, Kumar A, Neve RL, Nestler EJ. Sustained hippocampal chromatin regulation in a mouse model of depression and antidepressant action. Nat Neurosci. 2006;9:519-25.

[60] Covington HE 3rd, Maze I, LaPlant QC, Vialou VF, Ohnishi YN, Berton O, et al. Antidepressant actions of histone deacetylase inhibitors. J Neurosci. 2009;29:11451-11460.

[61] Hollis F, Wang H, Dietz D, Gunjan A, Kabbaj M. The effects of repeated social defeat on long-term depressive-like behavior and shortterm histone modifications in the hippocampus in male Sprague-Dawley rats. Psychopharmacology (Berl). 2010;211:69-77.

[62] Elliott E, Ezra-Nevo G, Regev L, Neufeld-Cohen A, Chen A. Resilience to social stress coincides with functional DNA methylation of the Crf gene in adult mice. Nat Neurosci. 2010;13:1351-1353.

[63] Uchida S, Hara K, Kobayashi A, Otsuki K, Yamagata H, Hobara T, et al. Epigenetic status of Gdnf in the ventral striatum determines susceptibility and adaptation to daily stressful events. Neuron. 2011;69:359-372.

[64] Champagne FA. Epigenetic mechanisms and the transgenerational effects of maternal care. Front Neuroendocrinol. 2008;29:386-397.

[65] Champagne FA, Meaney MJ. Stress during gestation alters postpartum maternal care and the development of the offspring in a rodent model. Biol Psychiatry. 2006;59:1227-1235.

[66] Champagne FA, Meaney MJ. Transgenerational effects of social environment on variations in maternal care and behavioral response to novelty. Behav Neurosci. 2007;121:1353-1363.

[67] Roth TL, Lubin FD, Funk AJ, Sweatt JD. Lasting epigenetic influence of early-life adversity on the BDNF gene. Biol Psychiatry. 2009;65:760-769.

[68] Weaver JR, Susiarjo M, Bartolomei MS. Imprinting and epigenetic changes in the early embryo. Mamm Genome. 2009;20:532-543.

[69] Anway MD, Cupp AS, Uzumcu M, Skinner MK. Epigenetic transgenerational actions of endocrine disruptors and male fertility. Science. 2005;308:1466-1469.

[70] Franklin TB, Russig H, Weiss IC, Graff J, Linder N, Michalon A, et al. Epigenetic transmission of the impact of early stress across generations. Biol Psychiatry. 2010;68:408-415.

[71] Wilkinson R, Pickett K. The Spirit Level: Why Equality is Better for Everyone. London, UK: Penguin; 2010.

[72] Marmot M. The Health Gap: the Challenge of an Unequal World. London, UK: Bloomsbury Publishing; 2015.

[73] Bourdieu P. The Forms Of Capital. In: Handbook Of Theory And Research For The Sociology Of Education. Greenwood, New York, NY, USA, 241-258 (1986).

[74] Coleman J. Foundations of Social Theory. Cambridge, USA: Harvard University Press; 1994.

[75] Putnam RD. Bowling Alone: The Collapse and Revival of American Community. NY, USA: Simon and Schuster; 2001.

[76] Dickerson SS, Kemeny ME. Acute stressors and cortisol responses: a theoretical integration and synthesis of laboratory research. Psychol Bull. 2004;130(3):355-391.

[77] Gilbert P, McEwan K, Bellew R, Mills A, Gale C. The dark side of competition: how competitive behaviour and striving to avoid inferiority are linked to depression, anxiety, stress and self-harm. Psychol Psychother. 2009;82(Pt 2):123-136.

[78] Layte R. The association between income inequality and mental health: testing status anxiety, social capital, and neo-materialist explanations. Eur Sociol Rev. 2011;28(4):498-511.

[79] Layte R, Whelan CT. Who feels inferior? A test of the status anxiety hypothesis of social inequalities in health. Eur Sociol Rev. 2014;30(4): 525-535.

[80] Mascaro O, Csibra G. Representation of stable social dominance relations by

human infants. Proc Natl Acad Sci USA. 2012;109(18):6862-6867.

[81] Frankel D, Arbel T. Group formation by two-year olds. Int J Behav Dev. 1980;3(3):287-298.

[82] Dickerson SS, Gable SL, Irwin MR, Aziz N, Kemeny ME. Social-evaluative threat and proinflammatory cytokine regulation: an experimental laboratory investigation. Psychol Sci. 2009;20(10):1237-1244.

[83] Chiang JJ, Eisenberger NI, Seeman TE, Taylor SE. Negative and competitive social interactions are related to heightened proinflammatory cytokine activity. Proc Natl Acad Sci USA. 2012;109(6):1878-1882.

[84] Packard CJ, Bezlyak V, McLean JS, et al. Early life socioeconomic adversity is associated in adult life with chronic inflammation, carotid atherosclerosis, poorer lung function and decreased cognitive performance: a cross-sectional, population-based study. BMC Public Health. 2011;11:42.

[85] Kumari M, Shipley M, Stafford M, Kivimaki M. Association of diurnal patterns in salivary cortisol with all-cause and cardiovascular mortality: findings from the Whitehall II study. J Clin Endocrinol Metab. 2011;96(5): 1478-1485.

[86] Kiecolt-Glaser JK, Gouin JP, Hantsoo L. Close relationships, inflammation, and health. Neurosci Biobehav Rev. 2010;35(1):33-38.

[87] Aslund C, Starrin B, Nilsson KW. Social capital in relation to depression, musculoskeletal pain, and psychosomatic symptoms: a cross-sectional study of a large population-based cohort of Swedish adolescents. BMC Public Health. 2010;10:715.

[88] Aslund C, Starrin B, Nilsson KW. Psychosomatic symptoms and low psychological well-being in relation to employment status: the influence of social capital in a large cross-sectional study in Sweden. Int J Equity Health. 2014;13:22.

[89] Youm Y, Laumann EO, Ferraro KF, et al. Social network properties and self-rated health in later life: comparisons from the Korean social life, health, and aging project and the national social life, health and aging project. BMC Geriatr. 2014;14:102.

[90] Uphoff EP, Pickett KE, Cabieses B, Small N, Wright J. A systematic review of the relationships between social capital and socioeconomic inequalities in health: a contribution to understanding the psychosocial pathway of health inequalities. Int J Equity Health. 2013;12:54.

[91] Fisher PA, Gunnar MR, Chamberlain P, Reid JB. Preventive intervention for maltreated preschool children: impact on children's behavior, neuroendocrine activity, and foster parent functioning. J Am Acad Child Adolesc Psychiatry. 2000;39(11): 1356-1364.

[92] Evans GW, Kim P, Ting AH, Tesher HB, Shannis D. Cumulative risk, maternal responsiveness, and allostatic load among young adolescents. Dev Psychol. 2007;43(2):341-351.

[93] Chen E, Miller GE, Kobor MS, Cole SW. Maternal warmth buffers the effects of low early-life socioeconomic status on pro-inflammatory signaling in adulthood. Mol Psychiatry. 2011;16(7):729-737.

[94] Miller GE, Brody GH, Yu T, Chen E. A family-oriented psychosocial intervention reduces inflammation in

low-SES African American youth. Proc Natl Acad Sci USA. 2014;111(31):11287-11292.

[95] Hostinar CE, Sullivan RM, Gunnar MR. Psychobiological mechanisms underlying the social buffering of the hypothalamic-pituitary-adrenocortical axis: a review of animal models and human studies across development. Psychol Bull. 2014;140(1):256-282.

[96] Borghol N, Suderman M, McArdle W, et al. Associations with early-life socio-economic position in adult DNA methylation. Int J Epidemiol. 2012;41(1):62-74.

[97] Lam LL, Emberly E, Fraser HB, et al. Factors underlying variable DNA methylation in a human community cohort. Proc Natl Acad Sci USA. 2012;109 Suppl. 2:17253-17260.

[98] McGuinness D, McGlynn LM, Johnson PC, et al. Socioeconomic status is associated with epigenetic differences in the pSoBid cohort. Int J Epidemiol. 2012;41(1):151-160.

[99] Needham BL, Smith JA, Zhao W, et al. Life course socioeconomic status and DNA methylation in genes related to stress reactivity and inflammation: the multi-ethnic study of atherosclerosis. Epigenetics. 2015;10(10):958-969.

[100] Stringhini S, Polidoro S, Sacerdote C, et al. Life-course socioeconomic status and DNA methylation of genes regulating inflammation. Int J Epidemiol. 2015;44(4):1320-1330.

[101] Fryers T, Melzer D, Jenkins R, Brugha T. The distribution of the common mental disorders: social inequalities in Europe. Clin Pract Epidemiol Ment Health. 2005;1:14.

[102] Swartz JR, Hariri AR, Williamson DE. An epigenetic mechanism links socioeconomic status to changes in depression-related brain function in high-risk adolescents. Mol Psychiatry. 2016; https://doi.org/10.1038/mp.2016.82.; Epub ahead of print.

[103] Beach SR, Brody GH, Todorov AA, Gunter TD, Philibert RA. Methylation at SLC6A4 is linked to family history of child abuse: an examination of the Iowa Adoptee sample. Am J Med Genet B Neuropsychiatr Genet. 2010;153b(2): 710-3.

[104] Beach SR, Brody GH, Lei MK, Kim S, Cui J, Philibert RA. Is serotonin transporter genotype associated with epigenetic susceptibility or vulnerability? Examination of the impact of socioeconomic status risk on African American youth. Dev Psychopathol. 2014;26(2):289-304.

[105] Zhao J, Goldberg J, Bremner JD, Vaccarino V. Association between promoter methylation of serotonin transporter gene and depressive symptoms: a monozygotic twin study. Psychosom Med. 2013;75(6):523-529.

[106] Nikolova YS, Koenen KC, Galea S, et al. Beyond genotype: serotonin transporter epigenetic modification predicts human brain function. Nat Neurosci. 2014;17(9):1153-1155.

[107] Swartz JR, Knodt AR, Radtke SR, Hariri AR. A neural biomarker of psychological vulnerability to future life stress. Neuron. 2015;85(3): 505-511.

[108] Swartz JR, Williamson DE, Hariri AR. Developmental change in amygdala reactivity during adolescence: effects of family history of depression and stressful life events. Am J Psychiatry. 2015;172(3):276-283.

[109] Yehuda R, Flory JD, Bierer LM, et al. Lower methylation of glucocorticoid receptor gene promoter 1F in peripheral blood of veterans with post-traumatic stress disorder. Biol Psychiatry. 2015;77(4):356-364.

[110] Sadeh N, Spielberg JM, Logue MW, et al. SKA2 methylation is associated with decreased prefrontal cortical thickness and greater PTSD severity among trauma-exposed veterans. Mol Psychiatry. 2016;21(3):357-363.

[111] Boks MP, Rutten BP, Geuze E, et al. SKA2 methylation is involved in cortisol stress reactivity and predicts the development of post-traumatic stress disorder (PTSD) after military deployment. Neuropsychopharmacology. 2016;41(5):1350-1356.

[112] Guintivano J, Brown T, Newcomer A, et al. Identification and replication of a combined epigenetic and genetic biomarker predicting suicide and suicidal behaviors. Am J Psychiatry. 2014;171(12):1287-1296.

[113] Kaminsky Z, Wilcox HC, Eaton WW, et al. Epigenetic and genetic variation at SKA2 predict suicidal behavior and posttraumatic stress disorder. Transl Psychiatry. 2015;5:e627.

[114] Yehuda R, Daskalakis NP, Bierer LM, et al. Holocaust exposure induced intergenerational effects on FKBP5 methylation. Biol Psychiatry. 2016;80(5):372-380.

[115] Cunliffe VT. The epigenetic impacts of social stress: how does social adversity become biologically embedded? Epigenomics. 2016;8(12):1653-1669.

Section 3

Background and Mechanisms Governing Optogenetics

Chapter 6

Cyanobacterial Phytochromes in Optogenetics

Sivasankari Sivaprakasam, Vinoth Mani,
Nagalakshmi Balasubramaniyan
and David Ravindran Abraham

Abstract

Optogenetics initially used plant photoreceptors to monitor neural circuits, later it has expanded to include engineered plant photoreceptors. Recently photoreceptors from bacteria, algae and cyanobacteria have been used as an optogenetic tool. Bilin-based photoreceptors are common light-sensitive photoswitches in plants, algae, bacteria and cyanobacteria. Here we discuss the photoreceptors from cyanobacteria. Several new photoreceptors have been explored in cyanobacteria which are now proposed as cyanobacteriochrome. The domains in the cyanobacteriochrome, light-induced signaling transduction, photoconversion, are the most attractive features for the optogenetic system. The wider spectral feature of cyanobacteriochrome from UV to visible radiation makes it a light potential sensitive optogenetic tool. Besides, cyanobacterial phytochrome responses to yellow, orange and blue light have more application in optogenetics. This chapter summarizes the photoconversion, phototaxis, cell aggregation, cell signaling mediated by cyanobacteriochrome and cyanophytochrome. As there is a wide range of cyanobacteriochrome and its combination delivers a varied light-sensitive response. Besides coordination among cyanobacteriochromes in cell signaling reduces the engineering of photoreceptors for the optogenetic system.

Keywords: cyanobacteriochrome, cyanophytochrome, photoswitch, photoreceptor, cell signaling transduction

1. Introduction

Photoreceptors in cyanobacteria are diverse in their spectral character from ultraviolet to visible wavelength. Plant photoreceptors were widely used in optogenetics, but their responses to specific wavelengths need more revision. When compared to these photoreceptors cyanobacteriochromes (CBCRs) receive more attention as a versatile optogenetic tool. Several photoreceptors respond to a wide range of light, photoconversion ability and photoswitches for dual light are new approaches and powerful tools for optogenetics [1]. Engineering of these photoreceptors will develop more versatile CBCR to alleviate the conventional methods like mutation and recombination [2]. Optogenetics in mammalian tissue adopted far-red illumination and adjacent infra-red radiance to visualize and activate responses in the cell. The CBCRs with linear tetrapyr-role is very sensitive to red and far-red light. Utilization of these infra-red sensitive and red light responsive CBCRs raised their application in optogenetics. So far phytochro-mobilin is used in mammalian cells recently cyanobacterial phytochrome 1 (CPH1) has been applied in mammalian cells proven its benefit in synthetic biology [3].

1.1 Cyanobacteria

Cyanobacteria are evolutionarily ancient phototrophic Gram-negative bacteria widely distributed in terrestrial, freshwater and marine environments. They are oxygenic photosynthesizers having major photosynthetic pigment chlorophyll-a and light-harvesting pigments phycobiliproteins. They survive in many extreme environ-ments, such as hot and cold deserts, hot springs, and hypersaline environments [4].

1.2 Cyanobacteriochrome

Light is an important factor for their nutrition and growth, therefore, it has a multitude photosensory complex that responds to a wide array of illumination. Each chromophore is a response to a particular wavelength based on the incident light it changes the arrangement and composition of pigments in the photon captur-ing antenna. This rearrangement of pigments to the incident light is the process of complementary chromatic acclimation. Cyanobacteria possess phototaxis move-ments it means they can move towards or away from specific light. Photoreceptors in cyanobacteria are commonly referred to as CBCRs [5].

1.3 Phytochromes

Generally, Phytochromes are photoreceptors that have been found in plants, algae, and bacteria. These photoreceptors are broadly utilized in biosensors and optoge-netics to screen and regulate diverse intracellular cycles like phosphorylation, gene activation, degradation of protein and change of calcium ions [6].

1.4 Phytochromes from cyanobacteria

Phytochromes are photochromic photoreceptors, generally responding to red and far-red radiation in the visible spectrum. Bilin is the most important portion in the chromophore and it is distributed in three different forms. Phytochrome in plants made of phytochromobilin, whereas in cyanobacteria it is in the form of Phycocyanobilin. Further Phytochromes in plants, algae and cyanobacteria constitute linear tetrapyrrole biliverdin [7]. The chromophore part in plant phytochrome has cysteine at the N terminal site of the protein. The phytochrome in plants differs from cyanobacteria by having biliverdin in the chromophore part. Evolutionary develop-ment in cyanobacteria brings out cysteine linked with biliverdin in the GAF domain and formed as phycocyanobilin also referred to as phytochromobilin. The transforma-tion of phytochrome into CBCR is due to changes in the molecular level.

1.5 Phytochrome classification

Phytochromes were primarily arranged into three subfamilies dependent on the number of domains in their photosensory core module (PCM). Phytochrome has three domains in their core structure, for example, PHY - phytochrome-explicit area, PAS - Per-Arnt-Sim, and GAF - cGMP phosphodiesterase-adenylate cyclase-FhlA. Even though the amino acid groupings of these domains have a dissimilar sequence, their structures were similar. Further subfamilies are cyanobacterial phytochromes (Cph), lack an N-terminal PAS area, and CBCRs, which contain a solitary GAF domain [8]. The domain proteins of PAS, GAF and PHY were interconnected to form homo and heterodimers [9].

1.6 Features of cyanobacteriochrome

Phytochrome in plants and algae has the sensitivity to the different light spectrum. Plant phytochromes are sensitive to red radiance furthermore, it performs red and far-red photoreversible photocycle. The phytochrome with bilin photoreceptors in eukaryotic green algae and prokaryotic cyanobacteria are sensitive to the visible spectrum [10–12].

The CBCRs are photoreceptors involved in the regulation of phototaxis. The photoreceptors SyCcaS, SyPixJ1, TePixJ, AnPixJ, SyCikA are now proposed to be CBCRs due to the presence of chromophore binding GAF domain.

- The domain GAF is enough for photoconversion

- chromophore in GAF domain varies from phytochrome GAF

- The GAF domain binds to linear tetrapyrrole pigments like phycoviolobilin or phycocyanobilin

- The chromes are responsive to a wide range of light from ultraviolet to the red region

2. CBCR in cyanobacteria

2.1 AnPixJ

The cyanobacterial genomes of Anabaena and Nostoc harbor *PixJ* homologs, having chromophore-linked GAF domains and domain MCP. The PixJ-GAF domains of *Anabaena* and *Nostoc* were distinct from the blue-shifted complex of CBCR TePixJ and CBCR SyPixJ1 [13]. The four GAF domains of PixJ are continuously arranged in AnPixJ of *Anabaena* sp. PCC 7120 (**Figure 1A**) that possess reversible photoconversion between red (648 nm) Pr[AnPixJ] to green (543 nm) absorbing form Pg (AnPixJ) [14]. Acidic denaturation of AnPixJ in *Anabaena* sp. PCC 7120 affected the gliding motility of hormogonia and phototaxis.

2.2 SyCcaS

Chromatic acclimation is an adaptive mechanism in some cyanobacteria capable of modifying their photosynthetic system reaction to the incident radiance. The phycocyanin content in *Synechocystis* sp. PCC 6803 is chromatically synchronized under red and green-orange light. The cells irradiated with red light produced a higher quantity of phycocyanin [15] than the cells exposed to green-orange light. The red

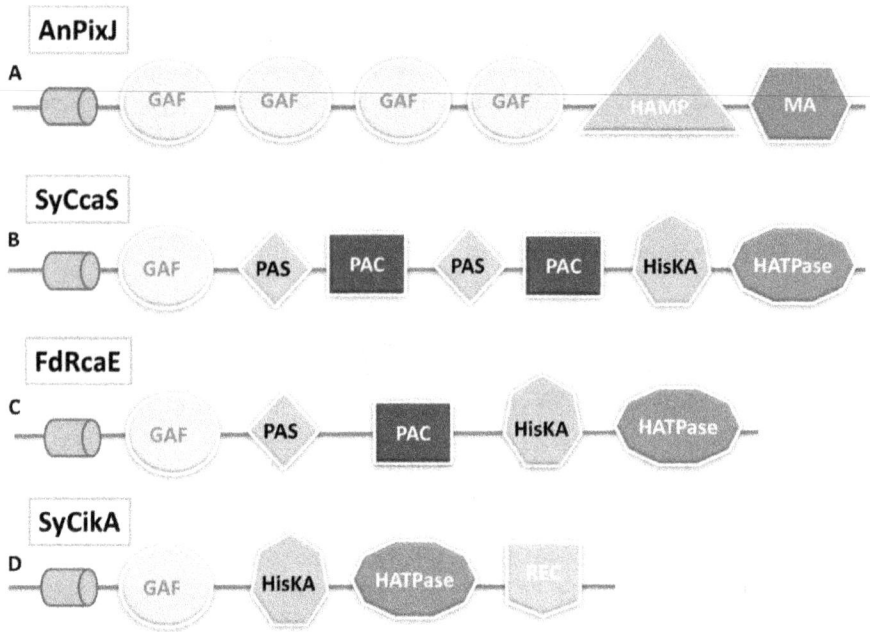

Figure 1.
Domain architecture of common cyanobacteriochromes (A) AnPixJ (B) SyCcaS (C) FdRcaE (D) SyCikA and their specific domains GAF with additional signaling domains are HAMP, methyl accepting chemotaxis protein (MA), PAS (PAS+ PAC- Photoswitchable adenyl cyclase), histidine kinase (HisKA+HATPase) and response regulator receiver domains (REC).

light condition activated the gene cpcG2 which encodes the synthesis of phycocyanin linker protein. Under red light CcaS, photoreceptor and transcriptional regulator CcaR induced the expression of the *cpcG2* gene [16]. It has a single GAF domain followed by PAS and PAC domains (**Figure 1B**).

2.3 FdRcaE

Fremyella diplosiphon harbor photoreceptor RcaE_GAF sector is homologous to SyCcaS_GAF. Genetic studies on FdRcaE revealed that it is a red light receptor, involved in the expression of operon *cpc2* encode synthesis of phycocyanin, [17] FdRcaE domain structure GAF, PAS and His kinase (**Figure 1C**), are parallel to SyCcaS (**Figure 1B**). Though the GAF domain is analogous to FdRcaE_ and SyCcaS, their light response is different in which the SyCcaS is a green light receptor. In *F. diplosiphon*, the green light has been used to activate genes for phycoerythrin post-translational modification and its linker polypeptides through the second signaling pathway by CBCR [18, 19]. The modern genome sequencing project would reveal the genetic background of the whole complementary chromatic acclimation process.

2.4 SyCikA

The chromophore-binding GAF domain of CikA in *Synechococcus elongatus sp.* PCC 7942, (**Figure 1D**) plays a crucial role in resetting the circadian rhythms [20].

Generally, the cyanobacterial chromophore is ligated with cysteine residue but it lacks the chromophore-tied Cys residue is parallel to other CikA homologs. Interconnection between the C-terminal pseudo-receiver domain and quinone is essential for the phase synchronizing of the rhythms [21]. CikA GAF domain of *Synechococcus* is comparable to the SyCikA_GAF of *Synechocystis* sp. PCC 6803. The properties of SyCikA are extremely uncommon however appear to be viable with the idea of circadian rhythms.

3. Functions of CBCR

3.1 Coordination of the cyanobacteriochromes

The photo biochemical properties of SesA holoprotein from the cyanobacterium *Thermosynechococcus vulcanus* have a blue light-responsive DGC (Diguanylate cyclase) activity. The SesB holoprotein isolated from *T. vulcanus* exhibited a reversible photoconversion system. It becomes blue light (417 nm) capturing form to a teal light (498 nm) assimilator. Another homologous CBCR from *T. vulcanus* is SesC which photoconverts a blue light (415 nm) assimilator to a green light (522 nm) absorber. These three CBCR proteins (SesA, SesB, and SesC) have phycoviobilin (PVB) and phycocyanobilin (PCB). These CBCR proteins were genetically expressed in *E. coli* which contains both PVB and phycocyanobilin [22, 23]. The SesA and SesB, perform independent photo conversion in *E. coli* in contrast when it is expressed in cyanobacteria it shows single photoconversion (**Figure 2**). Even though their spectral wavelength is different they coordinate and expressed single photocycle conversion.

SesB has GGDEF- type DGC (Diguanylate cyclase) domain (**Figure 3B**) and SesC has EAL- type PDE domain to deliver the c-di-GMP signal (**Figure 3C**). The SesB

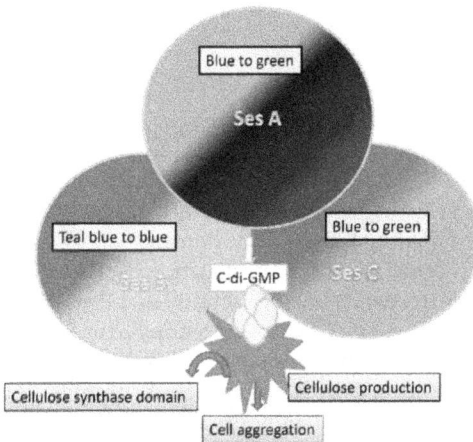

Figure 2.
Three different CBCR individually expressed to reveal different c-di-GMP signals. Ses A-produces c-di-GMP under blue light, Ses B- degrades c-di-GMP under teal light, Ses C- produces c-di-GMP under shorter wavelength and degraded c-di-GMP at the longer wavelength. These CBCR were coexpresses in Thermosynechococcus revealed c-di-GMP signal binds with cellulose synthase domain and promoted cell aggregation.

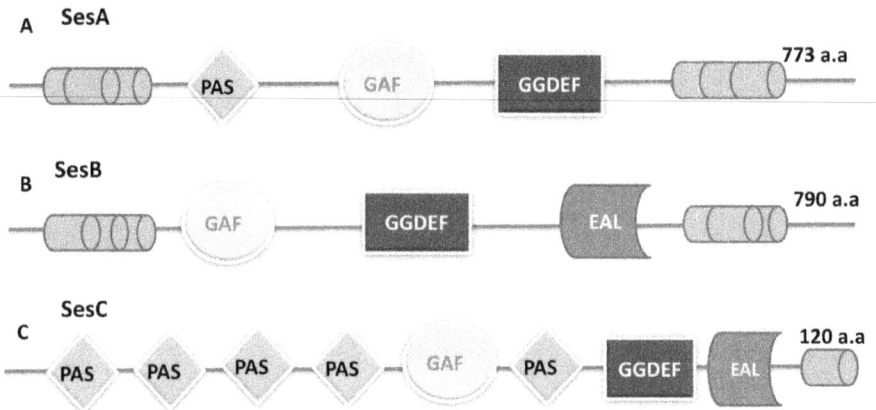

Figure 3.
Domain architecture of cyanobacteriochromes Ses A, Ses B and Ses C-GAF photosensitive domain and cell signaling domain PAS (Per/Arnt/Sim), GGDEF, EAL capped with a.a-amino acids.

DGC for c-di-GMP signal degraded under teal light, in contrast, expressed higher in blue light. In Ses, A c-di-GMP is higher under blue light and lowered in teal blue light. SesC DGF activity is maximum in blue light and minimum in green light. This is a chrome-responsive cyanobacterial c-di-GMP signaling coordination of (Ses –A, B and C) CBCRs.

  i. SesA a blue light-responsive DGC

  ii. SesB a teal light-responsive and GTP sensitive PDE

  iii. SesC, a dual-active CBCR having blue light-sensitive DGC and green light-responsive PDE activity

3.2 Cyanobacteriochrome in cell aggregation

The cell aggregation signaling molecule Cyclic dimeric guanosine monophosphate (c-di-GMP) is unique to cyanobacteria and bacteria [24]. Light is a key factor in controlling c-di-GMP signaling [25, 26]. The domain (GGDEF) for the synthesis and (EAL/HD-GYP) (**Figure 3A** and **B**) destruction of the c-di-GMP is higher in the CBCR GAF structure of freshwater cyanobacterial genomes. The CBCR induces the c-di-GMP signaling pathway. The CBCR—GAF domain of SesA (**Figure 3A**) from the thermophilic cyanobacterium *Thermosynechococcus elongatus* is activated by blue light irradiation, and disordering of *T. vulcanus* SesA inhibited cell aggregation.

Thermosyncechococcus spp., genomes possess five CBCR genes, three homologous CBCRs involved in the clumping of cyanobacterial cells are SesA (Tlr0924), SesB (Tlr1999), and SesC (Tlrtml). This CBCR has a photosensory domain with a c-di-GMP protein production/destruction domain. The CBCR-GAF domain of these three CBCR is involved in the light-controlled cell accumulation. There is a coordinated system of cell accumulation by c-di-GMP signaling via, Ses (A, B and C) CBCR (**Figure 2**) [27, 28].

3.3 Cyanobacterial photobiological responses

Prokaryotic photosynthetic organisms, cyanobacteria, depend on bilin-linked phytochromes (Cphs) and CBCRs, photoreceptors which are structurally and functionally vary from plant photoreceptors. The CBCRs are made of light-absorbing domains with various color-tuning and signal transmission processes, that make cyanobacteria capture a wide wavelength of light from UV–visible to far-red lights. The genome of filamentous cyanobacteria has a different type of CBCRs with wide chromophore-linked selectivity and photocycle protochromicity. The Cph lineage can absorb a wide range of light from blue-violet to yellow-orange light. This chapter also emphasized the color-sensitive diversity [29, 30] and signal transmission process of Cphs and CBCRs, concerning optogenetic.

Bilin-linked phytochrome Cphs and plant phytochromes (Phys) are similar in structure, with an N-terminal photosensory core module (PCM) and a C-terminal output regulatory module. The PCM contains the following domains PAS (Period/ Arnt/Single-minded), GAF(C-GMP phosphodiesterase/Adenylylcyclase/FhlA), and PHY (phytochrome-specific). The GAF domain is necessary for forming the bilin cross-linking; PAS and PHY structures are involved in bilin lyase activity [31]. Cyanobacteria have two types of bilin-linked photoreceptors Cphs, and CBCRs. In contrast to Cphs with PAS and PHY domain, CBCRs (lack PAS and PHY) absorb a wide array of light, by the GAF structure [32]. This wide array of light absorption by CBCR is called a color or spectral tuning mechanism.

3.4 CBCR in photobiological responses

Growth of the cyanobacterium *Synechocystis* PCC 6803 in red (R) and far red (FR) light is regulated by Cph1 and Cph2 in an antagonistic method. Modification in Cph1 negatively affects the *Synechocystis* growth in FR light, further destruction of Cph2 hinders its growth in red light [33]. Mutation in Cph2 transformed the growth rate and exopolysaccharide biofilm formation, involved in the control of the principal energy metabolism [34]. Under unusual light environments, the bilin conformation of the cyanobacterial antenna with light-absorbing phycobilisomes rearrangement is known as chromatic acclimation (CA). This process allows cyanobacteria to neutralize the proportion of light absorption between the photosystems [35, 36].

3.5 Dual light system

The CBCR response to two different light systems is mediated by the histidine (His) kinase domain. In *Leptolyngbya* sp. JSC-1, His domain is found in the proteins of Cph, RfpA, whereas CcaS in *Synechocystis* and *Nostoc punctiforme*, RcaE and DpxA in *Fremyella diplosiphon*, act as sensor kinase [35].

3.6 Phototaxis

The non-flagellated cyanobacteria adapt phototaxis in response to light. In *Synechocystis*, move towards light [37] and away from light [38] phototaxes are achieved by PixJ and UirS CBCRs. The CBCR- PixJ-GAF domain in *Synechococcus elongatus*, can respond to the direction of illumination by wavelengths that induce both progressive and refusal phototactic movements [39]. Other similar CBCR viz., SyPixJ [37], TePixJ [40], and AnPixJ [41] are commonly involved in phototaxis.

3.7 Photoinhibition

In some cyanobacteria, photoinhibition light conditions trigger the synthesis of photoprotective pigments. For example, intense radiation or UV radiation, accumulate mycosporine-like amino acids and scytonemins [30, 42]. Cyanobacteria, like *Nodularia* sp., *Euhalothece* sp. *Microcoleus* sp., and *Scytonema hofmanii* [43, 44] possess bilin photoreceptors, Cphs and CBCRs. These photosensitive receptors facilitate photobiological reactions by sensing and delivering signaling compounds.

3.8 Circadian clock

Cyanobacteria are responsive to diurnal photoperiods by adjusting their photosynthesis and respiration. In *S. elongatus* PCC 7942, the circadian clock controls the genes using promoters in light and dark conditions. Control of promoters is time-dependent, which sequentially maintains energy metabolism, cell division, and chromosome structure. Some CBCR domain (KaiABC), CikA (circadian input kinase A) and PsR in the *S. elongatus* oscillator become natural sensors that identify the change from light to dark by detecting the redox condition of the quinone pool [45].

3.9 Biofilm

Cyanobacteria form biofilms, which favor attachment on a surface to grow and produce extracellular polymers. This biofilm development in *Thermosynechococcus* is intervened by the cyclic diguanosine monophosphate (c-di-GMP) a bacterial secondary messenger [46]. Three CBCRs, SesA, B and C, in the blue/green light (ON/OFF) - c-di-GMP switch control non-motile and motile in planktonic networks [26, 47].

4. Photosensitive features of CBCR

4.1 Color sensing by Cphs and CBCRS

Cyanobacterial proteins contain the accompanying regions PAS-GAF-PHY [48]. Entire genome sequencing of cyanobacterial species, for example, *Microcoleus* IPAS B373 [49], *Euhalothece* Z-M001 [44], and *Tolypothrix* PCC7910 [50] are devoid of gene HY2, for phytochromobilin (PΦB) synthase. Further, these cyanobacteria have pcyA gene that encodes phycocyanobilin (PCB): ferredoxin oxidoreductase that catalyzes the conversion of biliverdin (BV) to PCB, a significant cofactor of Cphs and CBCRs [32, 51]. The quantity of Cphs and CBCRs differ among cyanobacteria, *Euhalothece* has 3 numbers, *Synechocystis* (8), *Microcoleus* IPAS B353 (9), *Acaryochloris marina* (12), *N. punctiforme* (18), and *Tolypothrix* PCC 7910 (36). In cyanobacterium, the complete number of bilin photoreceptors relies upon the size of its genome [49]. Besides, CBCRs are more plentiful in cyanobacteria, than Cphs, and the proportion of CBCRs for blue to red is corresponding to the environmental light conditions. For example, *Microcoleus* IPAS B353 grown in UV light developed only violet CBCRs than red/green and green/red CBCRs. Generally, UV light is recommended to develop and improve the quantity of short wavelength responsive CBCR [49].

4.2 Dual cays residues in CBCR for dual photocycle

Some CBCRs with exceptionally unchanged DXCF motif or the feebly rationed CXXR/K motif have extra Cys amino acids in the insertion loop (embed - Cys) via second thioether bond at the C10 atom under dark phase [52]. This sort of double Cys CBCRs, with a second thioether bond, is fragile and light-labile. These CBCRs are extremely responsive to capture violet or blue light in dark phase but it absorbs green, yellow, orange or greenish-blue light in the light phase. The cyanobacterial CBCRs are primarily linked to PCB yet some may link to phycoviolobilin (PVB) like Cphs [53, 54]. The change of PCB into PVB is unique to the DXCF-CBCRs subfamily [22]. The color tuning systems of CBCRs for far-red to orange (Fr/O) remain unidentified [55].

4.3 Signal transmission by CBCR

Cyanobacterial photoreceptors associated with signal transmission through phosphor transfer or c-di-GMP. Phosphorelay is a signal transmission process engaged with the autophosphorylation of His amino acid residue by His kinases, continued by phosphotransfer in association with reaction controllers. A film bound His kinase CBCR-UirS in *Synechocystis* accompanied with the reaction controller AraC family and UirR roles as a UV absorbing two-segment signaling framework [38]. Signaling in the chromophorylation process is regulated by the cystathionine beta-synthase (CBS) in the N-terminal of SesA. This in SesA can bind to ATP, ADP, and AMP which regulate the signaling process in chromophorylation.

4.4 Autolyase and autoisomerase in CBCR

Cyanobacterial photoreceptors are also called CBCRs that are similar to phytochromes [56]. PixJ GAF, from a thermophilic cyanobacterium *Thermosynechococcus elongatus*, regulates phototaxis. The BP-1 bacterial photoreceptors (TePixJ_GAF) reveal reversible photoconversion between a blue light (433 nm) capturer and a green light (531 nm) capturer. TePixJ GAF chromoprotein expressed in *Synechocystis* was denatured using acidic urea (8 M urea/HCl, pH 2.0) and it was compared with the cyanobacterial phytochrome Cphl having chromophore phycocyanobilin (PCB). The PCB is not a chromophore part in TePixJ, but PCB is a part of its isomer, phycoviolobilin (PVB). It confers the autolyase and autoisomerase property of GAF in TePixJ.

The primary CBCR for the phototaxis controller was recognized as PixJ. The CBCR SyPixJl of *Synechocystis* sp. PCC 6803 and TePixJ of *Thermosynechococcus elongatus* BP-1 showed selective reverse photo transfiguration between blue absorber (425-435 nm) Pb to green (531–535 nm) absorber Pg [57, 58]. Genetic modification in the pixJ of SypixJl and SypixD lost progressive phototaxis, these CBCR in original structure perceive blue light and characterize the order of motility as a regulatory switch [59]. The anticipated secondary arrangement of SyPixJl has N-terminal transmembrane helices, two successive GAF domains and a C-terminal methyl-accepting structure [5]. Proteolytic destruction and mass spectrometric investigation of SyPixJ 1_GAF and TePixJ_GAF showed that a straight tetrapyrrole was covalently bound to a peptide connected with phytochrome, a moderated Cys-His motif [5].

At the point when His6-TePixJ_GAF was digested with acidic urea in the dark phase, the Pb peak (433 nm) was changed from native form to a peak at 594 nm with a shoulder at 565 nm. The PVB in TePixJ_GAF captures a shorter wavelength of light than PCB. In any case, it ought to be noticed that the Pb absorb at 433 nm in native form is extraordinarily smaller than the urea-denatured PVB absorb at 594 nm. PVB is an isomer of PCB with a similar atomic mass, yet conjugated double bonds are detached at the C5 position. PVB with the apoprotein is accountable for the extraordinary blue capturing structure Pb and photoreversible modifications. The PCB transformation to PVB is because of the PecE and PecF proteins which are fundamental for ligation and isomerization.

5. Color-tuning mechanisms of cyanobacteriochromes

The term cyanobacteriochrome was first reported in 2004 by Dr. Ikeuchi in his paper about photoreceptor SypixJ1 [60]. This photoreceptor covalently binds a linear tetrapyrrole chromophore and performed reversible photoconversion between a blue-absorbing form (Pb) and a green-absorbing form (Pg). This protein has a cGMP-phosphodiesterase/adenylate cyclase/FhlA (GAF) similar to those of the Phytochromes. The GAF in cyanobacterial signals transduction proteins is identified as CBCRs. The GAF s of cyanobacteria are covalently linked to tetrapyrrole chromophore. This chromophore can sense different light (UV–visible spectra). In some cyanobacteria, this GAF regulate phototactic motility, chromatic acclimation and light-dependent cell aggregation.

5.1 CBCR structure and diversity

In CBCRs, the GAF is essential for chromophore incorporation and photoconversion [61] CBCRs have N and C terminals in which GAF s are located at the N-terminus. Signal output s viz. (HisKA + HATPase_c), Methyl-accepting (MA), GGDEF, and EAL s located at the C- terminus. His kinase is frequently detected as signal output s.

5.2 Chromophore variation in color-tuning mechanism

Generally, four types of linear tetrapyrrole chromophores, phycocyanobilin (PCB), biliverdin (BV), phytochromobilin (PFB) and phycoviolobilin (PVB), have been identified in the CBCR GAF domain. The mixture of these chromophores in the GAF domain results in broad and diverse spectral features [54, 62]. These chromophores are arranged in an order from a longer wavelength absorbing system to a shorter wavelength absorber (BV > PFB > PCB > PVB). The longer wavelength of light is absorbed by the longer length chromophores. Sometimes, the PCB isomerizes to PVB at the GAF region. This sort of PVB linked GAF area in CBCR has been distinguished from the cyanobacterium *Acaryochloris marina* [54]. The chromophore binding species should possess a UV-to-blue absorber.

The absorption of the Cys-free form is highly affected by the linked species of chromophores. The chromophore PCB in AM1_1186g2 revealed a reversible photoconversion between a Cys-free red-absorber (Pr) and a Cys-dependent blue absorber Pb [63], while covalently linked PVB in TePixJg showed reversible photoconversion between a Cys-free Pg absorber to a Cys-linked Pb absorber [62, 64]. The Cys-linked Pb absorber in the CBCR GAF s is the same as the blue to green reversible, but the Cys-free teal-absorber (Pt) is often shifted to blue absorber in association with the typical green absorber Pg. A twist in the D ring of the conserved Phe residues in the Pt absorber contributes to the blue-shift [65, 66]. The conversion of the Pt form into the typical red-shifted Pg form is mediated by the loss of these Phe residues [66]. Likewise in the XRG lineage, CBCR GAF s have red to green reversible photoconversion. Blue-shift of the Pg form to red absorbing Pr caused a small twist in the A and D ring [67]. This photocycle is mediated by the proton donor and the acceptor is Glu amino acid.

5.3 Dark reversion

The two distinctive light-harvesting types of the CBCR GAF domain are generally constant under the dark phase, thus these CBCR GAFs can detect the proportion of two wavelengths. Further, in some CBCR GAF showed unidirectional photoconversion and rapid dark reversion. This can identify the concentration of certain colors of light [68]. Some XRG CBCR GAF s are unidirectional photoconversion from Pr-to-Pg in dark conditions, and after 4–25 s it rapidly undergoes dark reversion from Pg-to-Pr [68]. The kinetics of these GAF s of the dark reversion is highly dependent on higher temperature [53, 69, 70]. Light and temperature were indulged in the regulation of these GAFs. These characteristics may be physiologically relevant to sense light intensity for efficient photosynthesis because the same light intensity with lower temperatures severely inhibited photosynthesis.

5.4 Engineering

Several CBCRs have been designed to change the color-tuning interaction and output activity. Inclusion of the second Cys residue and modifications of PCB-binding in the GAF s caused reversible photoconversion from red/green into blue/green. This is due to the isomerization of PCB to PVB by the incorporation of the second Cys residue which attaches to the chromophore in the reversible form [70, 71]. The twisted geometry of the D ring can also be removed [66]. The output activity of the native GAF was modified with other lineage s by adenylate cyclases [72–74] that respond to various light. Changing the length of the CBCR GAF linker region in CcaS and HisKA turns the light receptiveness of the green to red lineage [75].

6. Conclusions

Optogenetics, is a new branch of synthetic biology, is generally defined as the engineering of particular light-induced cellular reactions. This study was initiated with light-sensitive Phytochromes and bacteriochromes experimented in bacteria and mammalian neurons. Recently this research field is sensational due to the CBCRs from cyanobacteria are widely used as signaling components. These CBCRs in optogenetic systems performed the regulation of cellular responses spatially and temporally by precisely applying and removing light. A cyanobacterial photoreceptor-based optogenetic system was implemented to study the protein interaction and cell signaling in cyanobacteria, bacteria and mammalian cells. The application of CBCRs in optogenetic systems extends their usage in developing potential new therapeutics. Smaller size photosensory regions and autocatalytic activity of CBCRs are more advantageous than other photoreceptors.

Acknowledgements

The author S. Sivasankari thanks the DST-Government of India for financial support under the grant number DST/INSPIRE FELLOWSHIP/2014/IF140666.

Conflicts of interest

The author declares that there is no conflict of interest.

Abbreviation

CBCRs	Cyanobacteriochromes
CPH1	Cyanobacterial phytochrome 1
PCM	Photosensory core module
PAS	Per-ARNT-Sim
GAF	cGMP phosphodiesterase-adenylate cyclase-FhlA
PHY	Phytochrome-specific
Cph	Cyanobacterial Phytochromes
DGC	Diguanylate cyclase)
Phys	Phytochromes
UV	Ultraviolet
nm	Nanometer
Cys	Cysteine
PCB	Phycocyanobilin
PVB	Phycoviobilin
(PΦB)	Phytochromobilin
PFB	Phytochromobilin
Glu	Glutamine
Phe	Phenylalanine
BV	Biliverdin
MA	Methyl acceptor
ATP	Adenosine triphosphate
ADP	Adenosine diphosphate
AMP	Adenosine monophosphate
CBS	Cystathionine beta-synthase
His	Histidine

Cyanobacterial Phytochromes in Optogenetics
DOI: http://dx.doi.org/10.5772/intechopen.97522

Author details

Sivasankari Sivaprakasam[1*], Vinoth Mani[2], Nagalakshmi Balasubramaniyan[3] and David Ravindran Abraham[3]

1 Department of Microbiology, Jawaharlal Nehru College for Women, Thiruvalluvar University, Tamil Nadu, India

2 Department of Botany, Sri Vijay Vidhyalaya College for Arts and Science, Periyar University, Tamil Nadu, India

3 Department of Biology, Gandhigram Rural Institute-Deemed to be University, Tamil Nadu, India

*Address all correspondence to: san.s41190@gmail.com

IntechOpen

103

References

[1] Fushimi K, Narikawa R. Cyanobacteriochromes: photoreceptors covering the entire UV-to-visible spectrum. Current opinion in structural biology. 2019;57:39-46. DOI:10.1016/j.sbi.2019.01.018

[2] Bedbrook CN, Yang KK, Robinson JE, Mackey ED, Gradinaru V, Arnold FH. Machine learning-guided channelrhodopsin engineering enables minimally invasive optogenetics. Nature methods. 2019;16:1176-1184. DOI:10.1038/s41592-019-0583-8

[3] Müller K, Engesser R, Timmer J, Nagy F, Zurbriggen MD, Weber W. Synthesis of phycocyanobilin in mammalian cells. Chemical communications. 2013;49:8970-8972. DOI:10.1039/C3CC45065A

[4] Ho MY, Soulier NT, Canniffe DP, Shen G, Bryant DA. Light regulation of pigment and photosystem biosynthesis in cyanobacteria. Current opinion in plant biology. 2017;37:24-33. DOI:10.1016/j.pbi.2017.03.006

[5] Ikeuchi M, Ishizuka T. Cyanobacteriochromes: a new superfamily of tetrapyrrole-binding photoreceptors in cyanobacteria. Photochemical &Photobiological Sciences. 2008;7:1159-1167. DOI:10.1039/B802660M

[6] Chernov KG, Redchuk TA, Omelina ES, Verkhusha VV. Near-Infrared Fluorescent Proteins, Biosensors, and Optogenetic Tools Engineered from Phytochromes. Chem Rev. 2017; 117:6423-6446.

[7] Piatkevich KD, Subach FV, VerkhushaVV. Engineering of bacterial phytochromes for near-infrared imaging, sensing and light-control in mammals. ChemSoc Rev. 2013; 21;42:3441-3452.

[8] Giraud E, Verméglio A. Bacteriophytochromes in anoxygenic photosynthetic bacteria. Photosynth. Res. 2008;97:141-153. DOI: 10.1007/s11120-008-9323-0.

[9] Anders K, Essen LO. The family of phytochrome-like photoreceptors. Diverse,complex and multi- colored, but very useful. Curr. Opin. Struct. Biol. 2015;35:7-16. DOI: 10.1016/j.sbi.2015.07.005.

[10] Takala H, Bjorling A, Linna M, Westenhoff S, Ihalainen JA. Light-induced changes in the dimerization interface of bacteriophytochromes. J. Biol. Chem. 2015; 290:16383-16392. DOI: 10.1074/jbc.M115.650127.

[11] Takala H, Björling A, Berntsson O, Lehtivuori H, Niebling S, Hoernke M, Kosheleva I, Henning R, Menzel A, Ihalainen J.A., et al. Signal amplification and transduction in phytochromephotosensors. Nature. 2014;509:245-248. DOI: 10.1038/nature13310.

[12] Bellini D, Papiz MZ. Structure of a bacteriophytochrome and light-stimulated protomer swapping with a gene repressor. Structure. 2012;20:1436-1446. DOI: 10.1016/j.str.2012.06.002.

[13] Nultsch W, Schuchart H, Höhl M. Investigations on the phototactic orientation of Anabaena variabilis. Archives of Microbiology. 1979;122:85-91. DOI:10.1007/BF00408050

[14] Narikawa R, Fukushima Y, Ishizuka T, Itoh S, Ikeuchi M.A novel photoactive GAF of cyanobacterio-chromeAnPixJ that shows reversible

green/red photoconversion. Journal of molecular biology. 2008;25;38:844-855. DOI:10.1016/j.jmb.2008.05.035

[15] Sivasankari S, Vinoth M, Ravindran D, Baskar K, Alqarawi AA, Abd_Allah EF. Efficacy of red light for enhanced cell disruption and fluorescence intensity of phycocyanin. Bioprocess and Biosystems Engineering.2020;4:1-0. DOI:10.1007/s00449-020-02430-5

[16] Ikeuchi M, Ishizuka T. Cyanobacteriochromes: a new superfamily of tetrapyrrole-binding photoreceptors in cyanobacteria. Photochemical &Photobiological Sciences. 2008;7:1159-67. D OI:10:1039/B802660M

[17] Seib LO, Kehoe DM. A turquoise mutant genetically separates expression of genes encoding phycoerythrin and its associated linker peptides. Journal of Bacteriology. 2002;15;184:962-970. DOI: 10.1128/jb.184.1.962-970.2002

[18] Li L, Kehoe DM. In vivo analysis of the roles of conserved aspartate and histidine residues within a complex response regulator. Molecular microbiology. 2005;55:1538-1552. DOI: 10.1111/j.1365-2958.2005.04491.x

[19] Li L, Alvey RM, Bezy RP, Kehoe DM. Inverse transcriptional activities during complementary chromatic adaptation are controlled by the response regulator RcaC binding to red and green light responsive promoters. Molecular microbiology. 2008;68:286-297. DOI: 10.1111/j.1365-2958.2008.06151.x

[20] Schmitz O, Katayama M, Williams SB, Kondo T and Golden SS. CikA, a bacteriophytochrome that resets the cyanobacterial circadian clock, Science, 2000, 289, 765-768. DOI: 10.1126/science.289.5480.765

[21] Ivleva NB, Gao T, LiWang AC and Golden SS. Quinone sensing by the circadian input kinase of the cyanobacterial circadian clock, Proc. Natl. Acad. Sci. USA, 2006, 103, 17468-17473. DOI: 10.1073/pnas.0606639103

[22] Rockwell NC, Martin SS, Gulevich AG, Lagarias JC. Phycoviolobilin formation and spectral tuning in the DXCF cyanobacterio-chrome subfamily. Biochemistry. 2012; 21;51:1449-1463. DOI: 10.1021/bi201783j

[23] Enomoto G, Hirose Y, Narikawa R, Ikeuchi M. Thiol-based photocycle of the blue and teal light-sensing cyanobacteriochrome Tlr1999. Biochemistry. 2012; 10;51:3050-8. DOI:10.1021/bi300020u

[24] Hirose Y, Shimada T, Narikawa R, Katayama M, Ikeuchi M. Cyanobacterio-chromeCcaS is the green light receptor that induces the expression of phycobilisome linker protein. Proceedings of the National Academy of Sciences. 2008;15;105:9528-9533. DOI:10.1073/pnas. 0801826105

[25] Savakis P, De Causmaecker S, Angerer V, Ruppert U, Anders K, Essen LO, Wilde A. Light-induced alteration of c-di-GMP level controls motility of Synechocystis sp. PCC 6803. Molecular microbiology. 2012;85:239-251. DOI: 10.1111/j.1365-2958.2012.08106.x

[26] Enomoto G, Nomura R, Shimada T, Narikawa R, Ikeuchi M. CyanobacteriochromeSesA is a diguanylatecyclase that induces cell aggregation in Thermosynechococcus. Journal of Biological Chemistry. 2014;5;289(36):24801-9. DOI: 10.1074/jbc.M114.583674

[27] Barends TR, Hartmann E, Griese JJ, Beitlich T, Kirienko NV, Ryjenkov DA, Reinstein J, Shoeman RL, Gomelsky M,

Schlichting I. Structure and mechanism of a bacterial light-regulated cyclic nucleotide phosphodiesterase. Nature. 2009;459(7249):1015-1018. DOI:10.1038/nature07966

[28] Agostoni M, Koestler BJ, Waters CM, Williams BL, Montgomery BL. Occurrence of cyclic di-GMP-modulating output s in cyanobacteria: an illuminating perspective. MBio. 2013;30;4. DOI: 10.1128/mBio.00451-13

[29] Vinoth M, Sivasankari S, Ahamed AK, Al-Arjani AB, Abd_ Allah EF, Baskar K. Biological soil crust (BSC) is an effective biofertilizer on Vignamungo (L.). Saudi Journal of Biological Sciences. 2020;272325-32. DOI:10.1016/j.sjbs.2020.04.022

[30] Hirose Y, Rockwell NC, Nishiyama K, Narikawa R, Ukaji Y, Inomata K, Lagarias JC, Ikeuchi M. Green/red cyanobacteriochromes regulate complementary chromatic acclimation via a protochromicphotocycle. Proceedings of the National Academy of Sciences. 2013; 26;110:4974-4979. DOI: 10.1073/pnas.1302909110

[31] Rockwell NC, Lagarias JC. Phytochrome diversification in cyanobacteria and eukaryotic algae. Current opinion in plant biology. 2017;37:87-93. DOI:10.1016/j.pbi.2017.04.003

[32] Fushimi K, Narikawa R. Cyanobacteriochromes: photoreceptors covering the entire UV-to-visible spectrum. Current opinion in structural biology. 2019;1;57:39-46. DOI:10.1016/j.sbi.2019.01.018

[33] Fledler B, Broc D, Schubert H, Rediger A, Börner T, Wilde A. Involvement of Cyanobacterial Phytochromes in Growth Under Different Light Qualitities and

Quantities. Photochemistry and photobiology. 2004;79:551-555. DOI:10.1111/j.1751-1097.2004.tb01275.x

[34] Schwarzkopf M, Yoo YC, Hückelhoven R, Park YM, Proels RK. Cyanobacterial phytochrome2 regulates the heterotrophic metabolism and has a function in the heat and high-light stress response. Plant physiology. 2014;164:2157-2166. DOI:10.1104/pp.113.233270

[35] Sanfilippo JE, Garczarek L, Partensky F, Kehoe DM. Chromatic acclimation in cyanobacteria: a diverse and widespread process for optimizing photosynthesis. Annual review of microbiology. 2019;73:407-433. DOI:10.1104/pp.113.233270

[36] Wiltbank LB, Kehoe DM. Diverse light responses of cyanobacteria mediated by phytochrome superfamily photoreceptors. Nature Reviews Microbiology. 2019;17:37-50. DOI:10.1038/s41579-018-0110-4

[37] Yoshihara S, Ikeuchi M. Phototactic motility in the unicellular cyanobacteriumSynechocystis sp. PCC 6803. Photochemical &Photobiological Sciences. 2004;3:512-518. DOI:10.1039/B402320J

[38] Song JY, Cho HS, Cho JI, Jeon JS, Lagarias JC, Park YI. Near-UV cyanobacteriochromesignaling system elicits negative phototaxis in the cyanobacteriumSynechocystis sp. PCC 6803. Proceedings of the National Academy of Sciences. 2011;28;108: 10780-10785. DOI:10.1073/pnas.1104242108

[39] Yang Y, Lam V, Adomako M, Simkovsky R, Jakob A, Rockwell NC, Cohen SE, Taton A, Wang J, Lagarias JC, Wilde A. Phototaxis in a wild isolate of the cyanobacterium

Synechococcuselongatus. Proceedings of the National Academy of Sciences. 2018;26;115:E12378- 87. DOI:10.1073/pnas.1812871115

[40] Ishizuka T, Shimada T, Okajima K, Yoshihara S, Ochiai Y, Katayama M, Ikeuchi M. Characterization of cyano bacteriochrome TePixJ from a thermo philiccyanobacterium Thermo synecho coccuselongatus strain BP-1. Plant and cell physiology. 2006;1;47:1251-1261. DOI:10.1093/pcp/pcj095

[41] Narikawa R, Fukushima Y, Ishizuka T, Itoh S, Ikeuchi M. A novel photoactive GAF of cyanobacteriochromeAnPixJ that shows reversible green/red photoconversion. Journal of molecular biology. 2008;25;380:844-55. DOI:10.1016/j.jmb.2008.05.035

[42] Rastogi RP, Sinha RP, Moh SH, Lee TK, Kottuparambil S, Kim YJ, Rhee JS, Choi EM, Brown MT, Häder DP, Han T. Ultraviolet radiation and cyanobacteria. Journal of Photochemistry and Photobiology B: Biology. 2014;1;141:154-169. DOI:10.1016/j.jphotobiol.2014.09.020

[43] Sinha RP, Häder DP. UV-protectants in cyanobacteria. Plant Science. 2008;1;174(3):278-89. DOI:10.1016/j.plantsci.2007.12.004

[44] Yang HW, Song JY, Cho SM, Kwon HC, Pan CH, Park YI. Genomic survey of salt acclimation-related genes in the halophilic Cyanobacterium euhalothece sp. Z-M001.Scientific reports. 2020;20;10(1):1-1.DOI:10.1038/s41598-020-57546-1

[45] Cohen SE, Golden SS. Circadian rhythms in cyanobacteria. Microbiology and Molecular Biology Reviews. 2015;1;79(4):373-85. DOI:10.1128/MMBR.00036-15

[46] Agostoni M, Waters CM, Montgomery BL. Regulation of biofilm formation and cellular buoyancy through modulating intracellular cyclic di-cGMP levels in engineered cyanobacteria. Biotechnology and bioengineering. 2016;113(2):311-319. DOI:10.1002/bit.25712

[47] Enomoto G, Narikawa R, Ikeuchi M. Three cyanobacteriochromes work together to form a light color-sensitive input system for c-di-GMP signaling of cell aggregation. Proceedings of the National Academy of Sciences. 2015;30;112:8082-8087. DOI:10.1073/pnas.1504228112

[48] Yeh KC, Wu SH, Murphy JT, Lagarias JC. A cyanobacterialphyto-chrome two-component light sensory system. Science. 1997;5;277(5331):1505-8. DOI:10.1126/science.277.5331.1505

[49] Cho SM, Jeoung SC, Song JY, Kupriyanova EV, Pronina NA, Lee BW, Jo SW, Park BS, Choi SB, Song JJ, Park YI. Genomic survey and biochemical analysis of recombinant candidate cyanobacteriochromes reveals enrichment for near UV/violet sensors in the halotolerant and alkaliphiliccyanobacteriumMicrocoleus IPPAS B353. Journal of Biological Chemistry. 2015;20;290:28502-28514. DOI: 10.1074/jbc.M115.669150

[50] Song JY, Lee HY, Yang HW, Song JJ, Lagarias JC, Park YI. Spectral and photochemical diversity of tandem cysteine cyanobacterial phytochromes. Journal of Biological Chemistry. 2020;8;295:6754-6766. DOI:10.1074/jbc.RA120.012950

[51] Fujita Y, Tsujimoto R, Aoki R. Evolutionary aspects and regulation of tetrapyrrole biosynthesis in cyanobacteria under aerobic and

anaerobic environments. Life. 2015;5: 1172-1203. DOI:10.3390/life5021172

[52] Cho SM, Jeoung SC, Song JY, Song JJ, Park YI. Hydrophobic residues near the bilin chromophore-binding pocket modulate spectral tuning of insert-Cys subfamily cyanobacteriochromes. Scientific reports. 2017;17;7:1-2. DOI:10.1038/srep40576

[53] Fushimi K, Nakajima T, Aono Y, Yamamoto T, Ikeuchi M, Sato M, Narikawa R. Photoconversion and fluorescence properties of a red/green-type cyanobacteriochrome AM1_C0023g2 that binds not only phycocyanobilin but also biliverdin. Frontiers in microbiology. 2016;26;7:588. DOI:10.3389/fmicb.2016.00588

[54] Narikawa R, Nakajima T, Aono Y, Fushimi K, Enomoto G, Itoh S, Sato M, Ikeuchi M. A biliverdin-binding cyanobacteriochrome from the chlorophyll d–bearing cyanobacterium-Acaryochloris marina. Scientific reports. 2015;22;5:1-0. DOI:10.1038/srep07950

[55] Rockwell NC, Martin SS, Lagarias JC. Identification of cyanobacteriochromes detecting far-red light. Biochemistry. 2016;19;55:3907-3919. DOI:10.1021/acs.biochem.6b00299

[56] Ohmori M, Ikeuchi M, Sato N, Wolk P, Kaneko T, Ogawa T, Kanehisa M, Goto S, Kawashima S, Okamoto S, Yoshimura H. Characterization of genes encoding multi- proteins in the genome of the filamentous nitrogen-fixing cyanobacterium Anabaena sp. strain PCC 7120. DNAresearch. 2001;1;8:271-284. DOI:10.1093/dnares/8.6.271

[57] Yoshihara S, Katayama M, Geng X, Ikeuchi M. Cyanobacterialphytochrome-like PixJ1 holoprotein shows novel reversible photoconversion between blue-and green-absorbing forms. Plant

and Cell Physiology. 2004;15;45:1729-1737. DOI:10.1093/pcp/pch214

[58] Ishizuka T, Shimada T, Okajima K, Yoshihara S, Ochiai Y, Katayama M, Ikeuchi M. Characterization of cyano bacteriochrome TePixJ from a thermo philiccyanobacterium Thermosynecho coccuselongatus strain BP-1. Plant and cell physiology. 2006;1;47:1251-1261. DOI:10.1093/pcp/pcj095

[59] Gan F, Zhang S, Rockwell NC, Martin SS, Lagarias JC, Bryant DA. Extensive remodeling of a cyanobacterial photosynthetic apparatus infra-redlight. Science.2014;12;345(6202):1312-7. DOI: 10.1126/science.1256963

[60] Yoshihara S, Katayama M, Geng X, Ikeuchi M. Cyanobacterialphytochrome-like PixJ1 holoprotein shows novel reversible photoconversion between blue-and green-absorbing forms. Plant and Cell Physiology. 2004;15;45(12): 1729-37. DOI:10.1093/pcp/pch214

[61] Anders K, Essen LO. The family of phytochrome-like photoreceptors: diverse, complex and multi-colored, but very useful. Current Opinion in Structural Biology. 2015;1;35:7-16. DOI:10.1016/j.sbi.2015.07.005

[62] Ishizuka T, Narikawa R, Kohchi T, Katayama M, Ikeuchi M. CyanobacteriochromeTePixJ of Thermosynechococcus elongates harbors phycoviolobilin as a chromophore. Plant and cell physiology. 2007;1;48:1385-1390. DOI:10.1093/pcp/pcm106

[63] Narikawa R, Enomoto G, Fushimi K, Ikeuchi M. A new type of dual-Cyscyanobacteriochrome GAF found in cyanobacterium Acaryochloris marina, which has an unusual red/blue reversible photoconversion cycle. Biochemistry. 2014;12;53:5051-5059. DOI: 10.1021/bi500376b

[64] Ishizuka T, Kamiya A, Suzuki H, Narikawa R, Noguchi T, Kohchi T, Inomata K, Ikeuchi M. The cyanobacteriochrome, TePixJ, isomerizes its own chromophore by converting phycocyanobilin to phycoviolobilin. Biochemistry. 2011;15;50:953-961. DOI:10.1021/bi101626t

[65] Rockwell NC, Martin SS, Lagarias JC. Mechanistic insight into the photosensory versatility of DXCF cyanobacteriochromes. Biochemistry. 2012;1;51:3576-3585. DOI:10.1021/bi300171s

[66] Rockwell NC, Martin SS, Gulevich AG, Lagarias JC. Conserved phenylalanine residues are required for blue-shifting of cyanobacteriochrome photoproducts. Biochemistry. 2014;20;53:3118-3130. DOI:10.1021/bi500037a

[67] Lim S, Yu Q, Gottlieb SM, Chang CW, Rockwell NC, Martin SS, Madsen D, Lagarias JC, Larsen DS, Ames JB. Correlating structural and photochemical heterogeneity in cyanobacteriochrome NpR6012g4. Proceedings of the National Academy of Sciences. 2018;24;115:4387-4392. DOI:10.1073/pnas.1720682115

[68] Rockwell NC, Martin SS, Lagarias JC. Red/green cyanobacteriochromes: sensors of color and power. Biochemistry. 2012;4;51:9667-9677. DOI:10.1021/bi3013565

[69] Hasegawa M, Fushimi K, Miyake K, Nakajima T, Oikawa Y, Enomoto G, Sato M, Ikeuchi M, Narikawa R. Molecular characterization of DXCF cyanobacteriochromes from the cyanobacterium Acaryochloris marina identifies a blue-light power sensor. Journal of Biological Chemistry. 2018;2;293:1713-1727. DOI:10.1074/jbc.M117.816553

[70] Rockwell NC, Martin SS, Lagarias JC. There and Back Again: Loss and Reacquisition of Two Cys Photocycles in Cyanobacteriochromes. Photochemistry and photobiology. 2017;93(3):741-754. DOI:10.1111/php.12708

[71] Fushimi K, Ikeuchi M, Narikawa R. The expanded red/green cyanobacteriochrome lineage: An evolutionary hot spot. Photochemistry and photobiology. 2017;93:903-906. DOI:10.1111/php.12764

[72] Fushimi K, Enomoto G, Ikeuchi M, Narikawa R. Distinctive Properties of Dark Reversion Kinetics between Two Red/Green Type Cyanobacteriochromes and their Application in the Photoregulation of cAMP Synthesis. Photochemistry and photobiology. 2017;93:681-691. DOI:10.1111/php.12732

[73] Hu PP, Guo R, Zhou M, Gärtner W, Zhao KH. The Red/GreenSwitching GAF3 of Cyanobacteriochrome Slr1393 from Synechocystis sp. PCC6803 Regulates the Activity of an Adenylyl Cyclase. ChemBioChem. 2018;4;19:1887-1895. DOI:10.1002/cbic.201800323

[74] Blain-Hartung M, Rockwell NC, Moreno MV, Martin SS, Gan F, Bryant DA, Lagarias JC. Cyanobacteriochrome-based photoswitchable adenylyl cyclases (cPACs) for broad spectrum light regulation of cAMP levels in cells. Journal of Biological Chemistry. 2018;1;293:8473-8483. DOI:10.1074/jbc.RA118.002258

[75] Nakajima M, Ferri S, Rögner M, Sode K. Construction of a miniaturized chromatic acclimation sensor from cyanobacteria with reversed response to a light signal. Scientific reports. 2016;24;6:1-8. DOI: 10.1038/srep37595

Chapter 7

Functional Mechanism of Proton Pump-Type Rhodopsins Found in Various Microorganisms as a Potential Effective Tool in Optogenetics

Jun Tamogami and Takashi Kikukawa

Abstract

Microbial rhodopsins, which are photoreceptive membrane proteins consisting of seven α-helical structural apoproteins (opsin) and a covalently attached retinal chromophore, are one of the most frequently used optogenetic tools. Since the first success of neuronal activation by channelrhodopsin, various microbial rhodopsins functioning as ion channels or pumps have been applied to optogenetics. The use of light-driven ion pumps to generate large negative membrane potentials allows the silencing of neural activity. Although anion-conductive channelrhodopsins have been recently discovered, light-driven outward H^+-pumping rhodopsins, which can generate a larger photoinduced current than a light-driven inward Cl^--pump halorhodopsin, must be more efficient tools for this purpose and have been often utilized for optogenetics. There are abundant proton pumps in the microbial world, providing numerous candidates for potential practical optogenetic instruments. In addition, their distinctive features (that is, being accompanied by photoinduced intracellular pH changes) could enable expansion of this technique to versatile applications. Thus, intensive investigation of the molecular mechanisms of various microbial H^+-pumps may be useful for the exploration of more potent tools and the creation of effectively designed mutants. In this chapter, we focus on the functional mechanism of microbial H^+-pumping rhodopsins. Further, we describe the future prospects of these rhodopsins for optogenetic applications.

Keywords: Microbial rhodopsin, Photocycle, Proton transfer, Neural silencing, Optical pH control

1. Introduction

Optical control of biological reactions is one of the most recently studied fields of research because light facilitates highly spatial and temporal manipulation. In particular, optogenetics, that is, the specific and noninvasive control of biological activities such as neural activities by light stimulus of photoreceptor proteins heterogeneously expressed in targeted neurons or other related cells, has a significant impact in the field of neuroscience [1–8] and has attracted the interest of myriad researchers in the life sciences. Over the past 15 years since the first report on optogenetics in 2005 [1], the development of tools for this interesting technique has been rapidly progressing [9–14]. Recently, various types of photosensitive proteins have been employed for optogenetics [15–17]. Nevertheless, retinal-based proteins found in microbes (referred to as microbial rhodopsins), which were first applied to optogenetics, are still over-riding toolkits [18, 19].

Microbial rhodopsins (also termed type-I rhodopsins) are seven transmembrane α-helical proteins that bind to the retinal chromophore, similar to animal rhodopsins (also termed type-II rhodopsins) [20]. A distinctive property of animal rhodopsins is the difference in their chromophore configurations; retinals in microbial and animal rhodopsins adopt all-*trans* and 11-*cis* forms in the dark state, respectively. In addition, by all-*trans*-to-13-*cis* isomerization of the retinal with illumination, microbial rhodopsins undergo a linear cyclic photoreaction called photocycle, in contrast to animal rhodopsins, whose retinals are isolated from the protein moiety during their photoreaction processes. Their functions are also different; in addition to photo-sensing functions of animal rhodopsins, microbial rhodopsins also act as light dependent-ion transporters that can carry various types of ions such as H^+, Na^+, and Cl^- [21–24].

Microbial rhodopsins are classified into two categories of ion carriers. One is a light-gated ion channel, and the other is a light-driven ion pump. The former group includes channelrhodopsins (ChRs) [8, 25–27] and anion channelrhodopsins (ACRs) [28–30], which are the principal tools for optogenetics. Upon illumination, ChRs become permeable to various monovalent or divalent cations, such as H^+, K^+, Na^+, and Ca^{2+} [8, 25–27]. Therefore, in nerve cells expressing ChRs, the influx of Na^+ induced by light activation of ChRs causes depolarization in these cells, leading to neural activation [1–8, 25–27]. Conversely, light activation of ACRs, which act as anion-selective channels, can drive the hyperpolarization of ACR-expressing cells to suppress neural activity [28, 31]. The ion pump group includes light-driven outward H^+- [32, 33], Na^+- [34], and inward Cl^--pumps [35–38]. As these proteins can generate negative membrane potential in their incorporated cells by illumination, they can be utilized as neural silencers similar to ACRs [39, 40]. Microbial rhodopsins, as ion channels or pumps, can lead to changes in membrane potential by absorption of a photon without going through complicated reactions. This simple light-activated machinery makes them more easily applicable to optogenetics, together with repeatable properties through their photocycle.

Among the three types of ion-pumping rhodopsins, proton-pumping rhodopsins have a distinct feature from the other two. Proton translocation across the cell membrane induced by light activation of these pigments is accompanied by a change in intracellular pH. Hence, these proteins have the potential for various applications, for example, photoinduced pH control in cells or all sorts of organelles, as well as their use as neural silencers. To date, genes encoding H^+-pumping rhodopsins have been identified from the genomes of many microorganisms, irrespective of species [41], which enables us to gain the most plentiful genetic information from the database of the microbial rhodopsin family. Therefore, these types of rhodopsins may be applicable for exploring better candidates for optogenetics in various respects, such as the strength of neural inhibition, spectral properties (maximum absorption wavelength for activation), and kinetics.

Chow et al. screened efficient neuronal silencing rhodopsins and showed that the magnitude of photocurrents evoked by the activation of H^+-pump-type rhodopsins was on average higher than those evoked by the activation of inward Cl⁻-pump halorhodopsins (HRs) [39]. Moreover, the rates of activation upon light irradiation and recovery from inactivation after light cessation tended to be faster, as observed for archaerhodopsin-3 from *Halorubrum sodomense* (aR-3 or Arch), which is currently the most powerful H⁺-pumping tool for neural suppression, unlike HRs that retain long-lasting inactive states [39]. Based on these observations, H⁺-pump rhodopsins are considered more effective for the light-induced inhibition of neurons. Thus, these experimental facts for the practical use of H⁺-pumping rhodopsins have been steadily amassed; however, the utility of H⁺-pumping rhodopsins for optogenetics has not been completely evaluated from the molecular viewpoint. Therefore, an overview of the molecular mechanism of various H⁺-pumping rhodopsins, including newly found H⁺-pumps, may be useful for further development and rational design of optogenetic instruments. Here, we describe the functional mechanism of H⁺-pumping rhodopsins, particularly highlighting the aspect of photochemistry and the accompanying proton movement, with their future prospects for optogenetic applications.

2. H⁺-pumping rhodopsins from various microbial species

2.1 H⁺-pumps in archaebacteria

Among all microbial rhodopsins, the first H⁺-pumping rhodopsin reported was bacteriorhodopsin (BR), which was discovered in *Halobacterium salinarum* living in salt lakes or salterns in 1971 [42]. Haloarchaea, including those described above, can survive even in extremely salty environments with low oxygen concentrations using BR-based phototrophy, which is accomplished by ATP synthesis driven through a proton gradient produced with outward proton translocation across the cell membrane. Haloarchaeal plasma membranes contain deeply purplish patches (referred to as purple membranes), in which BR forms highly dense assemblies in the form of a two-dimensional hexagonal lattice. The high BR expression in native membranes, along with its highly stable property, facilitated biochemical and biophysical investigations of this protein by various approaches, including spectroscopic and structural methods [32, 33, 43–48]. Thus, BR is the most well-studied H⁺-pump.

Following the discovery of BR, the second H⁺-pump identified was archaerhodopsin (aR). Two homologous proteins, archaerhodopsin-1 and -2 (aR-1 and aR-2), were simultaneously identified from *Halobacterium* sp. aus-1 and aus-2 isolated from a lake in Western Australia by Mukohata et al. [49]. Several aR homologous proteins, including aR-3 described above, have been discovered in different haloarchaeal species [50–53]. In addition, Mukohata et al. successively identified two other H⁺-pump-like proteins belonging to a different clade from BR and aR: cruxrhodopsin-1 (cR-1) from *Haloarcula argentinensis* [54] and deltarhodopsin-1 (dR-1) from *Haloterrigena* sp. arg-4 [50]. Several homologs of these H⁺-pumps have also been identified in other species [55–57]. aRs, cRs, and dRs are very similar H⁺-pumps to BR; however, they are classified as apparently different tribes [50].

2.2 Eubacterial H⁺-pumps

The history of microbial rhodopsin research has been confined to the archaebacterial world for about three decades since the first discovery of BR. However, since the 2000s, rapid technical advances in metagenomics have led to the discovery of unknown microbial H⁺-pumping rhodopsins from various eubacteria [58, 59]. A representative example is proteorhodopsin (PR) from marine bacteria [60, 61].

In 2000, PR was first identified in the genome of uncultivated marine γ-proteobacteria, which is a member of the SAR86 clade, from a sea sample collected from Monterey Bay in California [62]. Thus, the nomenclature of this protein, i.e., "proteo-," originates from the name of the hosting bacterium. Sequencing of a bacterial artificial chromosome vector into which a fragmented DNA extracted from samples was cloned revealed the presence of a gene encoding rhodopsin-like protein (EBAC31A08) [62]. Furthermore, after transformation by this gene and successive induction of protein expression with exogenous retinal in *Escherichia coli*, acidification in suspension containing these PR-expressing cells was caused by illumination, indicating that PR can work as an outward light-driven BR-like H⁺-pump in the *E. coli* membrane [62]. After the first discovery of PR, further surveys demonstrated the existence of genes encoding novel PRs in not only γ-proteobacteria but also α-proteobacteria containing ubiquitous marine clades such as the SAR11 group [63], β-proteobacteria [64], and Flavobacteria [65, 66]. In addition, genes encoding numerous PR variants (>several hundreds or thousands of variants) have been identified in widespread oceans [67–72]. Nowadays, most marine bacterioplankton living in the photic zone are assumed to hold PR genes [41]. PR can be classified into two groups depending on their absorption maxima (λ_{max}): green-absorbing PR (GPR), whose λ_{max} is approximately 525 nm, and blue-absorbing PR (BPR) with a λ_{max} of ca. 490 nm [67, 69, 73–75]. The difference between these two groups is probably associated with the adaptation to the environments that the PR-retaining bacteria inhabit; most bacteria that are distributed at the surface of the sea and have access to available green light have GPR to obtain energy produced effectively using this wavelength of light, while bacteria at the depth of the sea water that exclusively have access to available blue light contain BPR [67, 74, 75].

PR-related proteins were also discovered from non-marine bacteria present in various environments, such as freshwater [76], high mountains [77], hot springs [78], and permafrost [79]. For example, a PR-like protein identified from actinobacteria living in freshwater is called actinorhodopsin (ActR) because it is classified into a phylogenetically different clade from PR [76]. A halophilic eubacterium *Salinibacter ruber* also contains a PR-like H⁺-pumping protein called xanthorhodopsin (XR) [80]. XR binds to the second chromophore, carotenoid salinixanthin, which acts as a light-harvesting antenna, expanding the spectral range for light activation of this protein because the energy obtained by light absorption of salinixanthin can be transferred to the retinal to induce isomerization [80, 81]. Another PR-like H⁺-pump with binding ability to salinixanthin, similar to XR [82], was discovered from the cyanobacterium *Gloeobacter violaceus* and called *Gloeobacter* rhodopsin (GR) [83]. Furthermore, a new type of H⁺-pump with a unique feature (described later) was discovered from a nonmarine gram-positive bacterium *Exiguobacterium sibiricum* present in Siberian permafrost samples, which was named *Exiguobacterium sibiricum* rhodopsin (ESR) [79]. Thus, PR-like eubacterial H⁺-pumping rhodopsins have been found in various archaea and bacteria [84, 85] and even in eukaryotic marine protists [86], which seems to have been achieved by lateral gene transfer [84].

2.3 Two types of H⁺-pumps from lower eukaryotes

In 1999, the presence of a gene encoding eukaryotic microbial rhodopsin (*nop-1*) was first found in the eukaryotic filamentous fungus *Neurospora crassa* [87]. This rhodopsin-like protein encoded by *nop-1* is called *Neurospora* rhodopsin (NR). The amino acid sequence of NR contained the requisite corresponding residues for proton pumping of BR; however, a previous photochemistry study using recombinant NR proteins heterogeneously expressed in the methylotrophic yeast *Pichia pastoris* revealed that NR showed a slower photocycle that is close to sensor-type rhodopsins [88]. Therefore, it is speculated that NR is physiologically associated with carotenoid biosynthesis regulation by functioning as a photosensor rather than a H^+-pump [89, 90], although its exact physiological role remains unknown. Later, other NR-related fungal opsin genes were discovered in a different fungal species, *Leptosphaeria maculans*, which is the fungal agent of blackleg in canola [91]. This opsin-coded protein is termed *Leptosphaeria* rhodopsin (LR or Mac). Through its characterization using proteins prepared by heterogeneous expression in yeast (*Pichia pastoris*) similar to NR, it was demonstrated that LR acts as a BR-like outward H^+-pump with a fast photocycle, unlike NR [91]. Furthermore, through advanced genomic analyses, new fungal rhodopsins that are classified into a third subgroup were identified. The fungal wheat pathogen *Phaeosphaeria nodorum* possesses two rhodopsin-like protein-encoding genes [92]. These fungal rhodopsins are called *Phaeosphaeria* rhodopsin 1 (PhaeoRD1) and *Phaeosphaeria* rhodopsin 2 (PhaeoRD2). PhaeoRD1 is an analogous protein to LR, whereas PhaeoRD2 is a member of the third group. Considering its coexistence with other rhodopsin forms from the same species, PhaeoRD2 is regarded as an auxiliary protein [92]. Characterization of these fungal rhodopsins heterogeneously expressed in *P. pastoris* suggested that both pigments exhibit fast photocycles that are characteristic of H^+-pump-type rhodopsins [92].

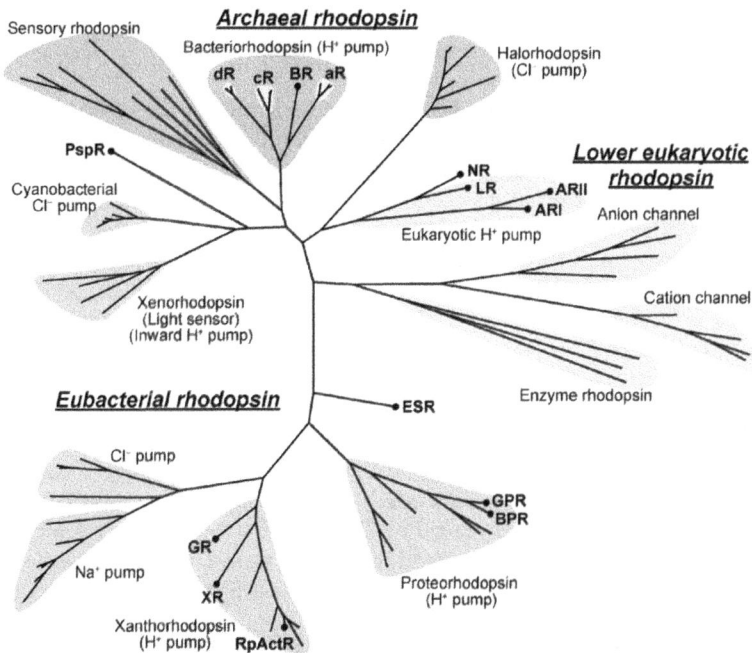

Figure 1.
Phylogenetic tree of microbial rhodopsins. RpActR represents ActR from actinobacterium Rhodoluna planktonica strain MWH-Dar1.

Acetabularia rhodopsin (AR) is another eukaryotic H^+-pump found in the giant unicellular marine green alga *Acetabularia acetabulum* [93]. *Acetabularia acetabulum*, which is also known as the "Mermaid's Wineglass", is an extremely interesting organism in terms of morphology because this unicell exhibits a unique complex life cycle comprising several distinct developmental phases [94]. In 2004, Mandoli et al. first reported the cDNA sequence of a fragmented possible opsin-encoding gene (*aop*) from juvenile *Acetabularia*. Subsequently, Hegemann et al. succeeded in cloning full-length opsin cDNA from this alga [93]. They heterogeneously expressed AR proteins in the membrane of *Xenopus laevis* oocytes and characterized the electrophysiological properties of this protein. Through a series of experiments, they demonstrated that AR is an outward light-driven H^+ pump [93]. Moreover, Jung et al. successively recloned two opsin genes from juvenile *Acetabularia*, which slightly differed from the gene cloned by Hegemann et al. The two AR homologs identified by them were named *Acetabularia* rhodopsin I and II (ARI and ARII, also abbreviated as Ace1 and Ace2, respectively) [95, 96]. Thus, two types of H^+-pumps from eukaryotic microorganisms are currently known: fungal and algal H^+-pumping rhodopsins (**Figure 1**).

3. Proton translocation mechanism of microbial H^+-pumping rhodopsins: from the photochemical and proton transfer viewpoints

3.1 Proton transport of BR: a typical model of H^+-pumping rhodopsins

When the molecular mechanism of microbial H^+-pumping rhodopsins is considered, the scenario of proton transportation in BR is often used as a prototype. Detailed descriptions of the H^+-pumping mechanism of BR from various aspects can be found in excellent previously published reviews (refer to relevant refs. [32, 33, 43-48]). We present only a brief outline here.

The photocycle of BR is initiated by photoisomerization of the retinal from all-*trans* to 13-*cis* upon formation of the K-intermediate. Then, during the transition between four sequentially formed photoproducts, L, M, N, and O intermediates, stepwise proton transfer reactions occur between amino acid residues buried within the protein or aqueous phases on both the cytoplasmic (CP) and extracellular (EC) sides. In these processes, three main groups play an essential role in proton transport. One is a part of the retinal Schiff base (SB), which represents a linkage with a specific lysine residue located at the center of the seventh helix of the protein (G-helix) ($Lys216^{BR}$, **Figure 2**). This portion is usually protonated in the unphotolyzed state (protonated retinal Schiff base, PSB). The other groups are two aspartic acid residues, $Asp85^{BR}$ and $Asp96^{BR}$, located in the EC and CP domains, respectively, on the C-helix. $Asp85^{BR}$ facilitates the first step of proton translocation upon the L–M transition as a proton acceptor from PSB, whereas $Asp96^{BR}$ works as a proton donor to deprotonated SB during M–N transition and is sequentially involved in proton uptake from the CP bulk upon N–O transition accompanied by 13-*cis*-to-all-*trans* retinal reisomerization. Both $Asp85^{BR}$ and $Asp96^{BR}$ are required for efficient proton pumps because substitutions of these residues with nonionizable residues abolished or significantly decreased H^+-pumping capability [97].

A proton releasing complex (PRC) comprising several internal H_2O and various residues on the EC surface such as $Tyr57^{BR}$, $Arg82^{BR}$, $Tyr83^{BR}$, $Ser193^{BR}$, $Glu194^{BR}$, $Glu204^{BR}$, and $Thr205^{BR}$ also participates in the proton transfer reaction of BR [98, 99], although it is not always an indispensable component for proton pumping.

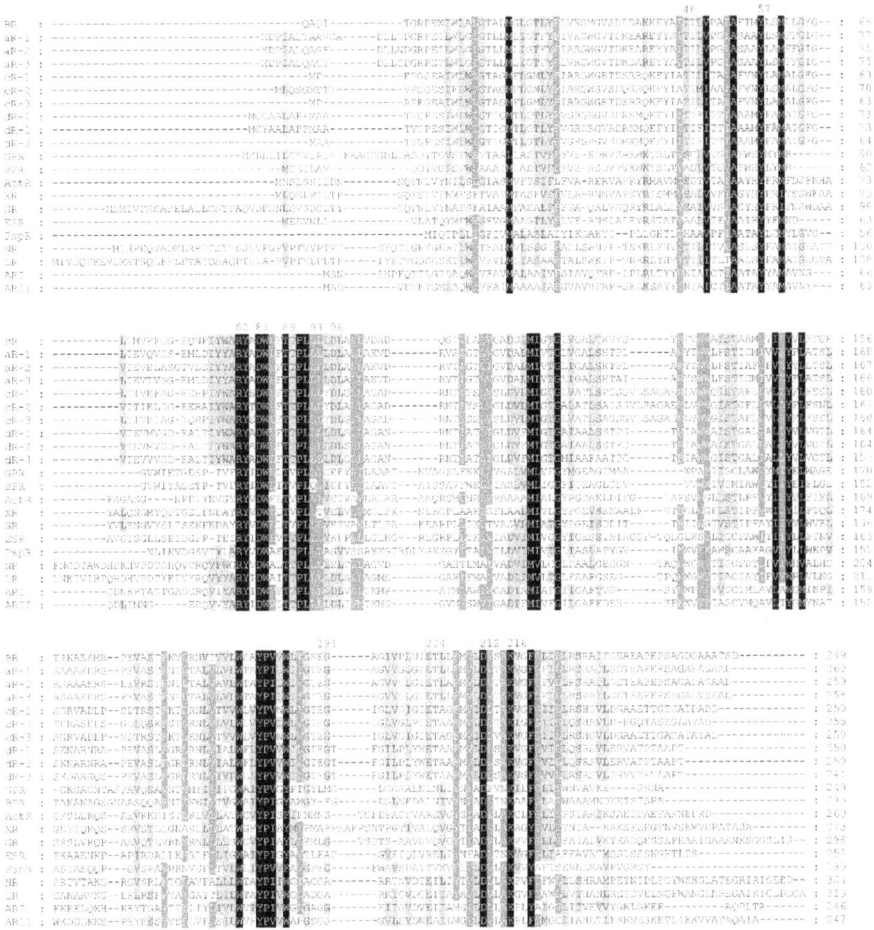

Figure 2.
Amino-acid alignment of various microbial H⁺-pumping rhodopsins. Analysis was performed using a multiple sequence alignment program (CLUSTALW). The numbers shown in the top row represent the numbering of amino acid residues in BR. The dotted line represents the missing residues in the determined structure. The amino acid residues with maximum homological numbers at each position are marked with a black or gray background depending on their numbers: The monochrome tone becomes darker as the number of homological residues increases. Notes: cR-2, cR from Haloarcula sp. arg-2; cR-3, cR from Haloarcula vallismortis; dR-2, dR from Haloterrigena turkmenica JCM9743; dR-3, dR from Haloterrigena thermotolerans; GPR, γ-proteobacterium (EBAC31A08) GPR; BPR, γ-proteobacterium (Hot75m4) BPR; ActR, RpActR.

The pK_a value of PRC in the H⁺-releasing M-state (~6) [100] divides the timing of proton release into two patterns: at pH values above ~6, a proton is initially released from PRC to the EC bulk during the L–M transition and the resultant deprotonated PRC receives a proton from the protonated Asp85^{BR} upon O-decay. In contrast, at pH values below ~6, the first such proton release from PRC upon M-rise does not occur, and a proton on PRC is released late upon O-decay [101].

Two threonine residues, Thr89^{BR} and Thr46^{BR}, are also important, although these residues do not belong to the series of proton transfer events due to nonionizable residues. Thr89^{BR} is within the active center and includes PSB, Asp85^{BR}, and some water molecules [102], where this residue forms a hydrogen bond with Asp85^{BR} [103], indirectly contributing to the initial proton transfer from PSB to Asp85[BR] during M-formation [102, 103]. In contrast, Thr46[BR] forms an interhelical hydrogen bond with Asp96[BR] in the CP region, which is associated with the regulation of pK_a in Asp96[BR] in the unphotolyzed state [104].

3.2 Common and different points on the amino-acid sequences among varying H[+]-pumps

In most outward H[+]-pumping microbial rhodopsins identified to date, the residues corresponding to three main groups (PSB [Lys216[BR] as the retinal binding site], Asp85[BR], and Asp96[BR]) described above are highly conserved. By checking their presence, we can therefore forecast whether each protein in the microbial rhodopsin family acts as an H[+]-pump like BR. **Figure 2** shows a comparison between important amino acid residues for proton transport among representative H[+]-pumping rhodopsins. As shown in this figure, almost all primal residues relevant to proton transport in archaeal-type H[+]-pumps agree with the residues corresponding to BR. Similarly, both fungal and algal H[+]-pumps from eukaryotes retain the residues corresponding to Asp85[BR] and Asp96[BR]; however, a difference exists in the components of PRC in BR. In both types of eukaryotic H[+]-pumps, the residue corresponding to Glu194[BR] of two EC glutamates in PRC is replaced by glycine, whereas another residue corresponding to Glu204[BR] is conserved. In contrast, in the eubacterial H[+]-pump, the residues corresponding to Asp96[BR] are substituted by conservative carboxylate glutamic acid, although there are several exceptions. Another significant aspartate corresponding to Asp85[BR] is perfectly conserved, similar to other types of H[+]-pumps. Furthermore, these H[+]-pumps lack both glutamic acids in the components of PRC: Glu194[BR] and Glu204[BR]. Thus, a comparison of the amino acid sequences among various H[+]-pumping rhodopsins can reveal the superconservation of the proton acceptor (Asp85[BR]) and the diversity of the proton donor (Asp96[BR]) and the residues in the EC proton releasing pathway. These differences could lead to different methods of proton transfer among varying H[+]-pumps.

3.3 Photocycles of other H[+]-pumping rhodopsins than BR

During a single photocycle induced by the absorption of one photon, ion-pump-type rhodopsins can transport ions as substrates. The number of photocycle turnover under illumination, therefore, affects the amount of ions transported by these proteins, in other words, the ion-pumping activity of these rhodopsins. In general, the turnover rate of the photocycle in ion-pumping rhodopsins tends to be relatively higher than those of photosensor-type rhodopsins to transport numerous ions per illumination. The speed of their photocycle completion can be used to analyze the H[+]-pump, in addition to actually measuring H[+]-pumping activity that is usually examined by measuring the photoinduced pH change in a suspension of cells expressing these rhodopsins. Furthermore, the identification of photointermediates during the photocycle of respective rhodopsins and the estimation of their rise/decay kinetics together with the measurement of transient proton transfer during their photocycles enable us to understand the timing of proton movement. Thus, detailed investigations of the photocycles are important for understanding the H[+]-pumping mechanism.

Among the H[+]-pumping rhodopsins identified so far, the next well-characterized proton pump following BR is GPR. In many studies, the first identified PR variant (EBAC31A08) was employed as a sample. As soon as GPR was discovered in 2000, various spectroscopic approaches such as static and time-resolved transient

UV–visible, FTIR, and FT-Raman spectroscopies were applied to characterize the photochemistry of this protein, as previously performed for the research of BR [105–110]. These experimental results revealed that the photocycle of GPR was similar to that of BR but also concomitantly contained several differences. Using the same kinetically analytical method as previously applied to the transient absorbance data of BR, where the possibilities of parallel or branch models were also considered [111], Váró et al. determined the photocycles of GPR at acidic and alkaline pH values [108, 109]. Their proposed photocycle at alkaline pH (9.5) is in accordance with the following scheme: GPR → K↔M$_1$ → M$_2$↔N↔GPR'(O) → GPR [108]. As shown in the above scheme, one of the apparent differences from the BR photocycle is the absence of L after K, which is thought to be probably due to kinetic reasons. A remarkably retarded (ca. 10-100-fold slower) decay of K compared with that of BR was observed in GPR [105]. Because of such slow K-decay, low-temperature Raman spectroscopic data presented by Fujisawa et al. demonstrated that the chromophore structure in GPR in the K state is less distorted compared to that of BR in the same state and is rather close to that of L in BR, which possess a more relaxed chromophore structure [112]. Therefore, the formation of a longer stable K state may obscure the appearance of L in the GPR along with the fast formation of the following M-state. Another difference from BR can be observed in the spectral characteristics of the latter photo-products, N and GPR'(O). The N-intermediate in PR was red-shifted with 13-*cis* chromophore retinal [105, 108] and resembled O in BR with respect to the absorption maxima. In addition, the GPR' intermediate had all-*trans* retinal chromophores similar to O in BR; however, its λ_{max} was very close to that of the original pigment. Friedrich et al. also determined the photocycle of GPR under both acidic (pH 5) and alkaline (pH 10) conditions based on a global fitting analysis (sequential irreversible model) for flash photolysis data [106]. In the latter half of the photocycle scheme proposed by them, after an equilibrium of M and a red-shifted O (λ_{max} = 580 nm) was produced, an equilibrium of N with a spectral property similar to that of the original pigment (λ_{max} = 530 nm) and O appeared [106]. Hence, if it is assumed that O and N in their scheme agree with N and GPR' in Váró's scheme, respectively, both schemes are compatible. The rate of photocycle turnover in GPR was fast (<several hundreds of milliseconds), although it was somewhat slower than that of BR (<several tens of milliseconds). In contrast, the photocycle of another type of PR, BPR, was slower by an order of magnitude than that of GPR [113]. The possibility of using BPR as a photosensor has been advocated, although its physiological role is still debated [113].

The photocycles of other eubacterial H$^+$-pumping rhodopsins, including XR, GR, ESR, and ActR, were also investigated by time-resolved absorption spectroscopy [80, 83, 114–116]. Their photocycles go through the K, L, M, N, and O states, similar to BR or GPR. For many eubacterial H$^+$-pumps including GPR, structural information obtained by multiple approaches such as X-ray crystallography, NMR, and atomic force microscopy has also been reported [117–123], providing structural insights into their photochemistry.

Recent genome analysis revealed that numerous eukaryotic fungi possess rhodopsin-like protein-encoding genes (RDs) and opsin-related genes (ORPs) [124]. Nevertheless, unlike archaeal or bacterial H$^+$-pumping rhodopsins, reports on the photochemistry of eukaryotic H$^+$-pumps are extremely limited because the protein expression in *Pichia pastoris* has been established only for a few fungal rhodopsins such as NR and LR. Meanwhile, several studies on the photochemical characterization of LR and its analogous protein PhaeoRD1 using visible and infrared spectroscopic techniques have been published [91, 92, 125–128]. These reports revealed that their photocycles include the K, L, M, N, and O states, similar to the BR photocycle [91, 92].

For two algal H$^+$-pumps, ARI and ARII, the establishment of a large-scale sample preparation method using a unique *Escherichia coli* cell-free membrane-protein production system developed by Shimono et al. [129] allowed the detailed elucidation of the spectroscopic and structural features of these proteins [96, 130–133]. Through global fitting analysis for time-dependent absorption changes based on a sequential irreversible model, we determined that the photocycles for ARI and ARII at near-neutral pH values can be represented by ARI → K↔L↔M↔N↔O → ARI' → ARI and ARII→K → L↔M↔N↔O → ARII' → ARII, respectively, which are very similar to the BR photocycle [130, 131]. However, the formation of a long M-N-O quasi-equilibrium was observed in both the photocycles of both AR proteins [130, 131], which is characteristic of these ARs. This indicates the presence of pronounced reverse reactions between M, N, and O in the photocycles of ARI and ARII. Although similar N–M or O–N back reactions also exist in BR, the rates of these reactions in BR are not significantly higher than those of ARI and ARII. The existence of prompt back reactions could hamper the fast turnover of the photocycle in these rhodopsins, thereby reducing H$^+$-pumping efficiency. However, a significantly faster O-rise and the irreversibility of the transitions from O to ARII' (a precursor of ARII) and from ARII' to the original state were observed during the photocycle of ARII [130]. Owing to these kinetic properties, the overall photocycle of ARII is a forward reaction, which may result in a turnover rate ($<\sim$100 ms at neutral pH) that is comparable to that of BR [130]. In contrast, the photocycle turnover of ARI was approximately 10-fold slower than that of ARII [131]. This may be attributed to slower decay of O and ARI' in the second half of the photocycle in ARI compared to the decay of O and ARII' in the photocycle of ARII.

3.4 Initial proton transfer from PSB to the proton acceptor, aspartate, upon L–M transition: the most crucial step for proton transport

As described above, the proton acceptor residue from PSB corresponding to Asp85^{BR} is superconserved in all H$^+$-pump-type rhodopsins, suggesting the significance of this residue in the proton pumping mechanism. The negative charge of deprotonated Asp85^{BR} interacts with another deprotonated aspartate Asp212^{BR} and three water molecules through hydrogen bonds, forming a pentagonal cluster that electrostatically stabilizes two positive charges of PSB and Arg82^{BR} [47]. The same cluster structure has also been observed in H$^+$-pumping rhodopsins other than BR [22, 131]. In this sense, two aspartates also play an important role in counterions to PSB, in which Asp85^{BR} and Asp212^{BR} are referred to as primary and secondary counterions, respectively. The aspartate residue, which is the proton acceptor, is deprotonated in the unphotolyzed state under physiological conditions. At pH values below the pK_a of the proton acceptor in the resting state, where this residue takes the protonated form, initial proton migration from PSB does not occur; thus, the formation of the M state with deprotonated SB is not observed and H$^+$-pumping activity vanishes. The pK_a of the proton acceptor in the unphotolyzed state, therefore, tends to adopt as low a value as possible, for example, ca. 2.5 for Asp85^{BR} [134]. Asp85^{BR} is conjugated with PRC located on the EC surface, which contributes to the retention of its low pK_a in the dark state [45, 134].

In contrast, the pK_a values of proton acceptor residues in eubacterial H$^+$-pumps tend to be relatively higher, for example, approx. 7-7.5 for GPR [105, 106, 109, 113], 7.8 (or 6.2) for BPR [113], 6.0 for XR [135], 4.5 for GR [136], 6.0 for ESR [137], and 5.8 for ActR [116]. Such high pK_a values in these pigments are thought to be associated with physiological pH conditions of the hosting bacteria possessing these rhodopsins; because the habitat of bacteria containing PR-like proteins (e.g., sea water, freshwater, etc.) usually has near-neutral or weakly alkaline pH conditions (approx. pH 6.5-8.5) that are above the pK_a values of the proton acceptors in the dark state, the proton acceptors of these H^+-pumps can adopt the deprotonated form to work as proton pumps. The elevated pK_a values of the proton acceptors in eubacterial-type H^+-pumps may be due to the absence of two EC glutamates corresponding to Glu194[BR] and Glu204[BR] and the repositioning of an arginine residue (corresponding to Arg82[BR]) located within the pentagonal cluster in the EC channel [138]. Moreover, in GPR, it was clarified that a highly conserved histidine residue His75[GPR] among bacterial H^+-pumps that is adjacent to the proton acceptor Asp97[GPR] contributes to the adjustment of the higher pK_a of Asp97[GPR] in the unphotolyzed state because the mutation of this residue significantly decreases the pK_a of Asp97[GPR] [139]. The replacement of the corresponding histidine residues in other eubacterial H^+-pumps, however, did not cause such a large change in the pK_a of their proton acceptor residues [136, 137], implying that the above-mentioned pK_a modulation mechanism through histidine is not common in all eubacterial-type H^+-pumps.

The primary and secondary counterions (corresponding to Asp85[BR] and Asp212[BR], respectively) are located near and arranged symmetrically around the PSB, resulting in forming a part of the proton acceptor cluster. The secondary counterion is also deprotonated like the primary counterion (proton acceptor) because the pK_a of this residue in the resting state usually takes a further lower value compared to the primary counterion. Nevertheless, a proton of PSB is always transferred to the primary counterion at the photoproduct rather than the secondary counterion. How should this proton transfer mechanism be considered? In the case of BR, it is thought that upon L–M transition, the pK_a of PSB is lowered from a value above ~13 in the dark state to a value below ~3, which is near the pK_a (~2.5 [134]) of Asp85[BR] in the same state. In contrast, the pK_a of Asp85[BR] simultaneously increases to a value of at least 8.5 approximately at this time (the first increase in the pK_a of Asp85[BR]) [140]. Thus, the pK_a values between PSB and Asp85[BR] are reversed, giving rise to a one-way proton movement from PSB to the deprotonated Asp85[BR]. Then, the pK_a of Asp85[BR] finally increases to above ~10 in the M-state, thus allowing it to maintain its protonated state until the end of the photocycle (a second increase in the pK_a of Asp85[BR]) [140]. The second increase in the pK_a of Asp85[BR] upon M-formation is thought to be triggered by the disruption of the electrostatic interaction between the negatively charged Asp85[BR] and the positively charged Arg82[BR] in PRC, which is caused by the protonation of Asp85[BR] and the accompanying deprotonation of PRC (initial proton release from PRC) during this process [140].

In contrast, a question that could arise would be how pK_a regulation in the proton acceptors of PR-like eubacterial H^+-pumps lacking their coupled PRC is achieved. Although there is no experimental evidence, we may presume that a similar pK_a inversion between PSB and its proton acceptor (Asp97[GPR]) in BR occurs upon the formation of M in GPR; the pK_a of PSB decreases from $> \sim 11$ in the unphotolyzed state [141] to ~3 upon M-rise, resulting in it being lower than the pK_a of Asp97[GPR] in the dark state (7-7.5). The possibility of an increase in the pK_a of Asp97[GPR] in the M-state similar to BR has also been reported [142]. FTIR data in DMPC-reconstituted vesicles revealed that the origin of the first proton release upon M-rise observed in GPR under alkaline conditions (pH ~9.5) is not Asp97[GPR], which is protonated during this transition [142]. This observation implies that the pK_a of Asp97[GPR] at M is above ~9.5. Why is the photoinduced pK_a increase in Asp97[GPR] caused by the absence of a

BR-like interaction with PRC? Although the reason is still unclear, an alternative interaction with neighboring His75^{GPR} [143] may work instead of the PRC, which is missing in GPR.

Through low-temperature FTIR experiments, it was suggested that PSB forms a stronger hydrogen bond with Asp227^{GPR} rather than Asp97^{GPR} within the pentagonal cluster around PSB upon K-formation [144]. In addition to this observation, the pK_a of Asp227^{GPR} in the unphotolyzed state was estimated to be approximately 2.6 or 3.0 [141, 145]. Hence, we cannot exclude the possibility that Asp227^{GPR} receives a proton from PSB at the photoproduct under such low pH conditions ($\sim$3 < pH < $\sim$7), where Asp97^{GPR} and Asp227^{GPR} are protonated and deprotonated at the resting state, respectively. Our experimental data using a rapid time-resolved pH-sensitive electrode method (described later with the details of this experimental method), however, showed that the pK_a of Asp227^{GPR} may further decrease from $\sim$3 at the dark state to $\sim$2.3 at the photolyzed state [145]. This possible pK_a decrease in Asp227^{GPR} at the photoproduct might hinder its proton acceptance from the PSB. Even though Asp227^{GPR} can transiently receive a proton from PSB, the proton might be immediately released to other dissociable residue(s) or internal waters. Interestingly, the computational calculations performed by Bondar et al. suggested that among three possible pathways of proton transfer from PSB to Asp85^{BR}, that is, 1) a direct pathway to Asp85^{BR} on the Thr89^{BR} side of the retinal, 2) a proton wire through Thr89^{BR}, and 3) a proton transfer pathway via Asp212^{BR}, the energy barrier of the third proton transfer pathway was the smallest [146]. Thus, a similar photoinduced pK_a decrease in the second counterion to GPR occurs even in other H$^+$-pumping rhodopsins, including BR, and might play a role in the initial proton movement from PSB to its proton acceptor upon light activation of these pigments. Further studies are required to clarify the roles of this mechanism.

3.5 Diverse proton transfer occurring on the CP side

Following the EC proton transfer in the first half of the photocycle, the CP proton transfer events via the SB proton donor in the second half of the photocycle after M-decay are the next critical steps. The proton transfer mechanism at this stage varies among the three types of H$^+$-pumping rhodopsins—archaeal, bacterial, and eukaryotic. In the latter half of the BR photocycle, the deprotonated SB first accepts a proton from its proton donor, Asp96^{BR}, located in the CP channel during the M–N transition. The pK_a of SB in this reprotonation process was estimated to be approximately 8 [147]. In contrast, the pK_a of Asp96^{BR} is maintained at a higher value (> $\sim$ 11) through an interhelical hydrogen bond with Thr46^{BR} on the B-helix [148]. Therefore, the pK_a of Asp96^{BR} needs to be lower than the pK_a value ($\sim$8) of SB to release a proton toward deprotonated SB, from > $\sim$ 11 at the initial state to $\sim$7-7.5 [149, 150]. This pK_a decrease is caused by the entry of water with the opening of the intracellular segment via the outward tilt of the F-helix at the M-state, leading to the internal hydration of the CP region. The inflow water breaks the interaction between Asp96^{BR} and Thr46^{BR}, facilitating hydrogen bonding rearrangement so that Asp96^{BR} forms a new interaction with neighboring water chains [151]. Then, during the following N–O transition, the pK_a value of Asp96^{BR} increases again and finally reaches a higher value (> $\sim$ 11) close to one in the dark state. Therefore, Asp96^{BR} can capture a proton from the CP medium to reprotonate.

As described above, in GPR, the residue corresponding to Asp96^{BR} is the conservative carboxylate, Glu108^{GPR}. This residue can function as a proton donor to SB;

however, the proton movement from Glu108^{GPR} to SB and the subsequent reprotonation of Glu108^{GPR} from the CP bulk are indistinguishable, unlike BR; two sequential proton transfer events in the CP channel concurrently take place upon the M–N transition [105]. The difference in CP proton migration in eubacterial H$^+$-pumps, including PR from BR, seems to be related to the difference in the environment around the proton donor in the intracellular part of the protein between them. In many eubacterial H$^+$-pumping rhodopsins, the interhelical hydrogen bonding pair corresponding to the Asp-Thr interaction in BR is replaced by the Glu-Ser interaction. The X-ray crystal structure of XR in the dark state revealed that the proton donor (Glu107^{XR}) in the CP channel connects to the peptide carbonyl of the lysine residue (Lys240^{XR}) in SB; therefore, the CP H-bonded chain via water is already formed in the unphotolyzed state [118]. Thus, the difference in the CP proton transfer scheme from BR may be due to the formation of the hydrophilic CP pathway in eubacterial H$^+$-pumps.

We also observed a further interesting characteristic in the CP proton transfer of the PR-like H$^+$-pump ESR. The residue positioned at the site of the proton donor in ESR is the cationic residue Lys96^{ESR} (see **Figure 2**). Nevertheless, Lys96^{ESR} seems to be involved in the CP proton transfer from the intracellular aqueous space to the inner deprotonated SB because the replacement of this residue by other nonionizable residues resulted in a significant delay of the M-intermediate [114]. This observation exploded a conventional concept, the so-called carboxyl rule, that the functional proton-donating residue is confined to two carboxylates (Asp or Glu). Some distinct structural features of BR can be observed in the X-ray crystal structure of the ESR. One of the differences is the presence of a cavity around Lys96^{ESR} located close to the CP bulk media [122]. Although Lys96^{ESR} is surrounded by hydrophobic residues in the CP channel in the dark state similar to BR, the cavity in the vicinity of Lys96^{ESR} is separated only by a polar side chain of Thr43^{ESR} (corresponding to Phe42^{BR}), in contrast to BR, whose proton donor residue is completely separated from the CP bulk solvent by a hydrophobic barrier composed of multiple hydrophobic residues including Phe42^{BR} [122]. Connectivity with the CP bulk facilitates direct access of the protons from the CP solvent in Lys96^{ESR}. Another difference is the flexibility of the side chain of Lys96^{ESR}, which may allow the smooth repositioning of this residue by donating to SB and reprotonation. Given that these structural properties are present in the CP region together with the time-resolved spectroscopic data using D$_2$O, it may be plausible that the CP proton transfer scheme in ESR is as follows [114]: Lys96^{ESR} adopts an unprotonated form at the resting state to be buried within the hydrophobic CP region. Upon M-decay, Lys96^{ESR} transiently catches a proton from the CP bulk solvent (at M$_1$↔M$_2$), and then, a little later, it donates a proton to SB (at M$_2$↔N$_1$). Hence, Lys96^{ESR} acts as a residue facilitating proton delivery from the CP bulk to the SB, which is an apparently different proton donating mechanism from the conventional one.

Another unique example of CP proton transfer was found in two types of gram-negative rod-shaped Proteobacteria in soil: *Pseudomonas putida* rhodopsin (PspR) from *Pseudomonas putida* and *Pantoea ananatis* rhodopsin (PaR) from *Pantoea ananatis*, a plant pathogen [152]. The notable properties of these types of rhodopsins are the replacement of the residue corresponding to Asp96^{BR} with nonionizable glycine and the presence of a specific histidine at the position corresponding to Thr46^{BR}. This histidine residue is highly conserved in a member of this group and is assumed to constitute a part of a proton-donating complex [152]. However, it was observed that the rate of M-decay linearly depends on the proton concentration of the medium in a

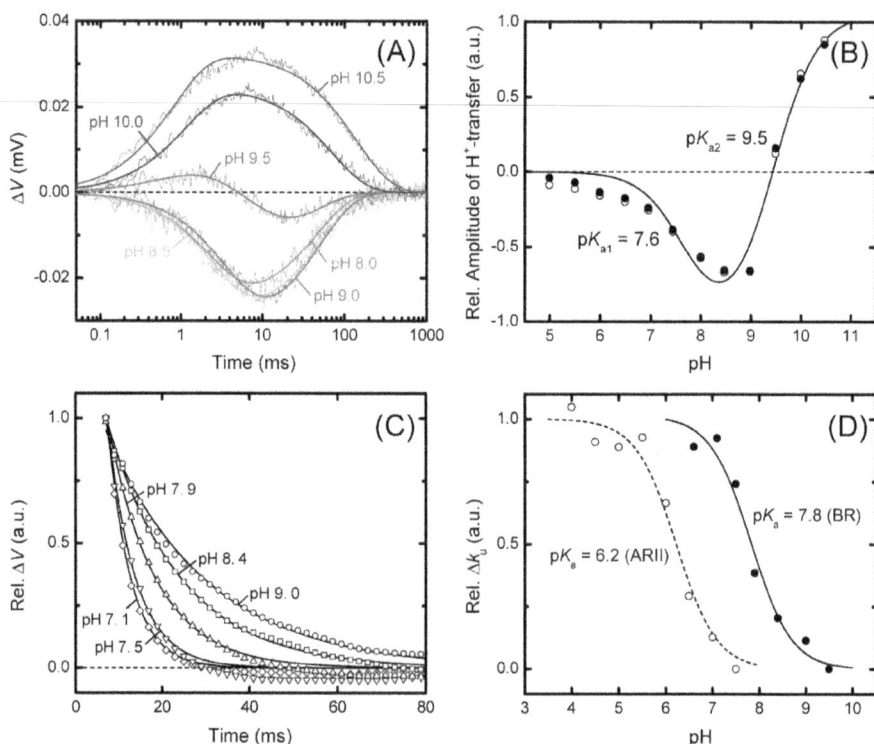

Figure 3.

pK_a estimation of critical residues for a proton pump by the SnO_2 (or ITO) electrode method. (A) Photoinduced voltage changes representing proton uptake/release at varying pH values [164]. Noisy and smooth curves represent the observed and fitted curves, respectively. For fitting, we employed the following kinetic equation:

$$\Delta Voltage \propto \Delta[H^+] = -A \frac{k_{f,u}}{k_{s,r}-k_{f,u}} \left(e^{-k_{f,u}t} - e^{-k_{s,r}t}\right) + B \frac{k_{f,r}}{k_{s,u}-k_{f,r}} \left(e^{-k_{f,r}t} - e^{-k_{s,u}t}\right),$$ where A represents a constant

reflecting the fraction of the subpopulation photoinducing the first proton uptake followed by release and the rate constants of the first H^+-uptake and second H^+-release in such a proton transfer sequence are expressed by $k_{f,u}$ and $k_{s,r}$, whereas B represents a constant reflecting the fraction of subpopulation inducing the opposite sequence of proton transfer, and the rate constants of the first H^+-release and second H^+-uptake in that case are expressed by $k_{f,r}$ and $k_{s,u}$, respectively. At pH < 9.5, where the first H^+-release cannot be obviously observed, it was assumed that B is almost zero. In contrast, the fitting at pH > 9.5 was conducted as $A = 0$. Six buffer agents with different pK_a values were added to the media for experiments so that the buffering action remained constant over a wide pH range ($\sim 5 \leq pH \leq \sim 11$). (B) Plot of amplitude of H^+-transfer versus pH. Filled and empty circles indicate plots of strict values with theoretical regression (-A and B values obtained by the above fitting) and approximate peak values of photoinduced signals estimated by sight, respectively. These values were plotted as relative values. A solid curve represents a curve fitted using the following equation: $\Delta[H^+] = -C\left(\frac{1}{1+10^{pK_{a1}-pH}}\right)\left[\left(\frac{1}{1+10^{pH-pK_{a2}}}\right) - \left(\frac{1}{1+10^{pK_{a2}-pH}}\right)\right]$,

where C, pK_{a1}, and pK_{a2} represent a scaling constant for the amplitude, pK_a values of $Asp97^{GPR}$, and an unidentified X-residue at the unphotolyzed state, respectively. The idea for the derivation of the equation has been described previously [164]. (C) Plots of the part of the second H^+-uptake after initial H^+-release as a function of time [162]. All values obtained at varying pH values ($\diamond$, pH 7.1; $\triangledown$, pH 7.5; $\triangle$, pH 7.9; $\square$, pH 8.4; $\bigcirc$, pH 9.0) were plotted as relative values. Continuous curves are fitting curves with single exponential eqs. (D) pH dependence of the rate constants of the second H^+-uptake (k_u). Increments of k_u at each pH obtained by subtraction of the minimum value at the highest pH were plotted as relative values. Filled and empty circles represent the plots for BR and ARII, respectively. These plots were well fitted with the Henderson–hasselbalch equation with a single pK_a value. Respective fitting curves for BR and ARII are shown using solid and broken curves. Panels A and B were adapted with permission from Tamogami et al. [164], Biochemistry copyright 2016 American Chemical Society, whereas panels C and D were adapted with permission from Tamogami et al. [162], Photochem. Photobiol. Copyright 2009 the authors, journal compilation, the American Society of Photobiology.

homologous protein in the same group [153], implying that the histidine forms a CP conductive channel rather than a proton-donating complex to enable rapid proton movement from the CP surface. Identification of the role of this unique histidine requires further study.

Among eukaryotic H^+-pumps, both fungal and algal H^+-pumps possess the same proton donor aspartate residue as BR. For two algal H^+-pump homologs ARI and ARII, however, the residue corresponding to $Thr46^{BR}$ is replaced by asparagine, which may cause different interactions with the proton donor and its pK_a regulation from BR and fungal-type rhodopsins. Our experimental data indeed revealed that the pK_a values of their proton donors ($Asp100^{ARI}$ and $Asp92^{ARII}$) in H^+-uptake (N-O transition) are ~6 (**Figure 3D**), which is ca. 1-1.5 units lower than that of BR (7-7.5) [130]. In the sensor-like fungal rhodopsin NR, the residue corresponding to $Asp96^{BR}$ is glutamic acid, similar to numerous eubacterial H^+-pumping rhodopsins, while the corresponding residue in the H^+-pump LR is aspartic acid, similar to BR. Interestingly, the substitution of the proton donor $Asp150^{LR}$ with an NR-like glutamate abolished the fast H^+-pumping photocycle [126, 127], implying that residues other than native aspartate work improperly in fungal H^+-pumps, even though it is a conservative one. In contrast, the influence of Asp-Glu replacement in $Asp96^{BR}$ differed depending on the experimental conditions [97, 154]. In the reconstituted BR heterogeneously expressed in *Escherichia coli*, the replaced glutamate residue fully functioned as a proton donor [97], whereas the replacement of BR in the native membrane led to a remarkable delay in SB reprotonation [154]. In contrast to the cases of BR or LR, the proton donating function of GPR was not lost by the substitution of $Glu108^{GPR}$ with BR and LR-like aspartate or even ESR-like lysine (data unpublished). Therefore, the distinct mechanisms of CP proton translocation via their proton donors and the specificity of the respective proton donors in the three types of H^+-pumping rhodopsins may originate from the difference in the environment around each proton donor in the CP channel.

3.6 Existence of two substates in the latter photoproducts of the photocycle and the chemical and structural events occurring during the transition between them

Among the three photointermediates M, N, and O produced in the latter half of the photocycle in H^+-pumping rhodopsins, two spectrally silent substates are known for each photoproduct [33, 132]. Because the transitions between these substates occur without apparent spectral changes, they are usually observed by kinetic analysis for transient absorbance changes measured using various spectroscopic techniques. Three critical events for proton translocation occur during these silent transitions. As is known in BR, the first crucial event was observed upon the transition between two successive M-states, M_1 and M_2, which is accompanied by the accessibility switch of SB from the EC side to the CP side. This switching is important for unidirectional proton transport because it causes the conversion of the direction of proton movement from toward EC at M to toward CP at N.

The second event occurs during the N_1-to-N_2 transition, where the accessibility of the proton donor changes. In BR, the proton donor $Asp96^{BR}$ connects to the SB but not the CP bulk during the M–N transition, thus hampering the misdirected transfer of a proton of $Asp96^{BR}$ toward the CP solvent. Then, the connection of $Asp96^{BR}$ to SB is switched toward the CP side upon the N_1–N_2 transition, facilitating the reprotonation of Asp96^{BR} from the CP surface [33, 45]. Although the detailed mechanism of this accessibility switch upon N$_1$–N$_2$ transition remains incompletely understood, even in the most well-known BR, a previous computational study by Wang et al. proposed a model in which the further opening of the proton uptake pathway in the CP channel, which remains closed even in the M-state with the opening of the F-helix by the presence of a hydrophobic barrier composed of Phe42^{BR} and multiple other hydrophobic residues, is triggered by the deprotonation of Asp96^{BR} during the M–N transition, leading to the connection of Asp96^{BR} to the CP aqueous space [155]. In contrast, for the algal H$^+$-pump ARII, it was presumed that the change in the unique interhelical interaction between Asp92^{ARII} and Cys218^{ARII} located in the CP domain acts as a switch for opening the gate of the CP channel for H$^+$-uptake [133].

In contrast to M and N, the molecular events in the O-state have not been completely examined because the stable trapping of O produced in the latter stages of the photocycle is difficult. In the early stages of studies on BR, Haupts et al. hypothesized that during the N–O transition, the reisomerization of the retinal from the 13-*cis* (15-*anti* PSB) to all-*trans* (15-*syn* PSB) form is followed by the switching of the N-H bond of PSB from the CP (15-*syn* PSB) to the EC (15-*anti* PSB) side, the so-called isomerization/switch/transfer (IST) model [156]. In contrast, the results of MD simulation performed by Wang et al. supported the opposite model (SIT model) as a more plausible scheme: the isomerization of the retinal from 13-*cis* (15-*syn* PSB) to all-*trans* (15-*anti* PSB) is preceded by the switching of PSB from the 15-*anti* to 15-*syn* forms [157]. If the scheme corresponds to the latter model, another substate with a 13-*cis* chromophore should be formed after the switching of PSB during the N$_2$-O transition with the thermal reisomerization of retinal. Thus, we attempted to detect the presence of further substates. In general, the existence of the quasi-equilibrium among M, N, and O states described above makes it difficult to observe O. However, in the algal H$^+$-pump ARII under acidic conditions (pH < ~5.5), N did not accumulate during the photocycle due to the presence of a rapid back reaction between M and N and the acceleration of proton uptake upon the following N–O transition under these conditions, resulting in notable observation of O [132]. Through kinetic analysis of time-resolved absorbance changes under these conditions, we succeeded in detecting two spectral analogous O-intermediates (O$_1$ and O$_2$) [132]. As the O$_1$–O$_2$ transition was accompanied by a faint but obvious red-shift of the absorption maximum, we assumed that the 13-*cis*-to-all-*trans* retinal isomerization occurs during the O$_1$–O$_2$ transition after the switching of PSB upon the N$_2$–O$_1$ transition based on the model proposed by Wang et al. [157]: O$_1$ is a precursor before the formation of O (O$_2$) with a twisted all-*trans* chromophore retinal. In previous studies on BR, it was reported that the steric contact of Lue93^{BR} with the 13-methyl group of retinal is significant for facilitated retinal reisomerization during this transition [158]. The residue corresponding to this leucine is almost completely conserved among all microbial rhodopsins (see **Figure 2**), implying that the mechanism described above is common in the microbial rhodopsin family.

3.7 Role of PRC formed on the EC surface

As described above, PRC located on the EC surface is not necessarily indispensable for proton pumping because of the presence of a PRC-deficient type (eubacterial or eukaryotic) H$^+$-pumping rhodopsins, although PRC alters the timing of proton release during the photocycle. The replacement of either Glu194^{BR}, Glu204^{BR}, or both by nonionizable residues, however, caused a delay in O-decay with a late proton release from the protonated Asp85[BR] toward the EC surface as well as the absence of the initial proton release upon the L–M transition. In addition, when the residues corresponding to three of PRC-constituting residues (Ser193[BR], Glu194[BR], and Thr205[BR]) in a sensory-type rhodopsin from *Natronomonas pharaonis* (NpSRII) were replaced by the same residues as BR, the lifetime of O in this triple NpSRII mutant became approximately 20-fold shorter than that of the wild type [159]. Hence, the presence of PRC on the EC surface may be involved in not only the early proton release toward the EC aqueous phase during the photocycle but also in the formation of a hydrophilic proton conductive pathway in the EC channel, which contributes to efficient proton translocation. In contrast, as observed in the X-ray crystal structure of ESR, there may be a cavity for the proton releasing pathway that already connects to the EC bulk solvent at the resting state in eubacterial H$^+$-pumping rhodopsins without PRC. Thus, in the EC domain of these H$^+$-pumping rhodopsins, a hydrophilic pathway may be formed in a different manner from archaeal H$^+$-pumps, participating in facilitated proton movement on the EC side.

3.8 Importance of the method for pK_a estimation of crucial residues involved in proton transfer

As described previously, sequential proton transfer events during the photocycles in various microbial H$^+$-pumping rhodopsins, including BR, are successfully accomplished by regulating rigorous pK_a changes among the crucial residues (particularly, PSB (or deprotonated SB), its proton acceptor, donor, and known or unknown proton-releasing residue(s)) related to proton translocation. The estimation of the pK_a values of these residues in both unphotolyzed and photolyzed states, therefore, provides important clues for understanding the proton transfer mechanism in these H$^+$-pumps. Such pK_a values can be indirectly estimated using spectroscopic approaches, such as FTIR or NMR. However, the establishment of a more direct method for pK_a estimation is preferable, which can be achieved by measuring the photoinduced proton exchange between the protein and media (proton uptake/release) arising as a result of proton transfer events during the photocycle at varying pH values.

As a method for measuring proton movement transiently occurring during the photocycles of these pigments, the conventional method of using various pH-indicator dyes is frequently employed [44]. This method is highly time-resolved because the transient pH changes of the media with photoinduced proton uptake and release in rhodopsins are monitored based on the real-time transient absorbance changes of these pH-sensitive dyes in the sample suspension. The use of this method, therefore, enables us to precisely identify the timing of proton uptake and release together with the rise and decay kinetics of photoproducts. However, the pH range for measurement is confined to the pH values around its pK_a; therefore, pK_a estimation using this method is difficult. In contrast, another method using a tin oxide (SnO$_2$ or indium-tin oxide, ITO) transparent electrode [160–163] is also highly pH-sensitive and rapidly time-resolved, although the applicable time period is limited within the time scale from several ten to hundred microseconds to hundreds of milliseconds [162, 164]. Moreover, this method can detect small pH changes with photoinduced proton uptake and release in the vicinity of a protein-attached electrode as a sufficiently large amplitude of voltage changes, even in solutions containing a small amount of buffer agents. Based on these advantages, we applied this method to the pK_a estimation in H$^+$-pumps from three biological kingdoms, BR, GPR, and ARII [130, 162, 164]. The pK_a estimation was performed in two ways. One method estimates the pK_a value from the pH dependence of the magnitude of voltage changes, reflecting proton uptake and release. **Figure 3A** shows the photoinduced transient proton transfers in GPR at varying pH values, where the upward and downward shifts represent proton release and uptake, respectively. Because the peak time and magnitude of these data depend on both the on- and off-time constants, the fraction of the subpopulation inducing proton uptake or release at the respective pH values is estimated by fitting with the kinetically derived theoretical equation. **Figure 3B** shows the plots of the values estimated from the above fitting analysis as a function of pH. From further fitting analysis for these plots with an equation developed based on the Henderson–Hasselbalch theory, we succeeded in estimating the pK_a values of some residues associated with photoinduced proton transfer events in GPR [164]. In contrast, the direct plots of amplitudes of light-induced proton transfer signals without strict regression described above, which are approximately proportional to the amount of proton transfer, also exhibited similar pH-dependent behavior, although these plots include some error. Therefore, such plotting may be useful as a method for simply estimating the approximate pK_a values. Another method estimates the pK_a values from the pH dependence of the kinetics of photoinduced proton uptake or release. **Figure 3C** shows the pH-dependent changes in the traces of the part representing the latter proton uptake following the initial release of a proton in BR in the pH range of 6.5 9.5. The fitting for these traces with a single exponential equation (solid curves in this figure) gave the rate constant values of proton uptake at the respective pH values. Similarly, the estimated values of the rate constant at each pH were plotted against pH (**Figure 3D**), and sequentially, the pK_a values of Asp96^{BR} in H$^+$-uptake (N-O transition) were estimated using the Henderson–Hasselbalch Equation [162]. All other BR values estimated by these methods were consistent with the corresponding values previously estimated using other experimental approaches [45] (also see **Figure 4B**). Therefore, this method for estimating the pK_a values of some crucial residues for proton pumping, which is an index of proton pumping efficiency, may be a powerful and effective tool for screening efficient H$^+$-pumps or their engineered mutants for optogenetics.

4. Future perspectives of H$^+$-pumping rhodopsins as optogenetic tools

As described at the beginning of this chapter, outward H$^+$-pumping microbial rhodopsins can evoke stronger light-induced neural suppression and quicker recovery from the inactivated state formed upon illumination than inward Cl$^-$-pump HRs. Hence, optical neural control using H$^+$-pumping rhodopsins may also be an effective alternative for optogenetics, although neural inhibition with light-gated anion channel ACRs has recently attracted attention. The rational design based on the functional molecular basis of these rhodopsins described in this chapter may allow the creation of "neo-type" H$^+$-pumping microbial rhodopsins by introducing several mutations to further enhance the effect of neural silencing upon illumination, resulting in the acceleration of the development of more efficient tools for optogenetics along with developing color variants with various spectral properties. Further increases in protein expression and stability in targeted neural cells could also lead to the improvement of optogenetic tools. For this purpose, taking advantage of the abundance of these H$^+$-pumping rhodopsins, the exploration of new microbial H$^+$-pumping rhodopsins with novel properties (e.g., high thermal stability [165, 166]) from nature may be useful for producing mutants.

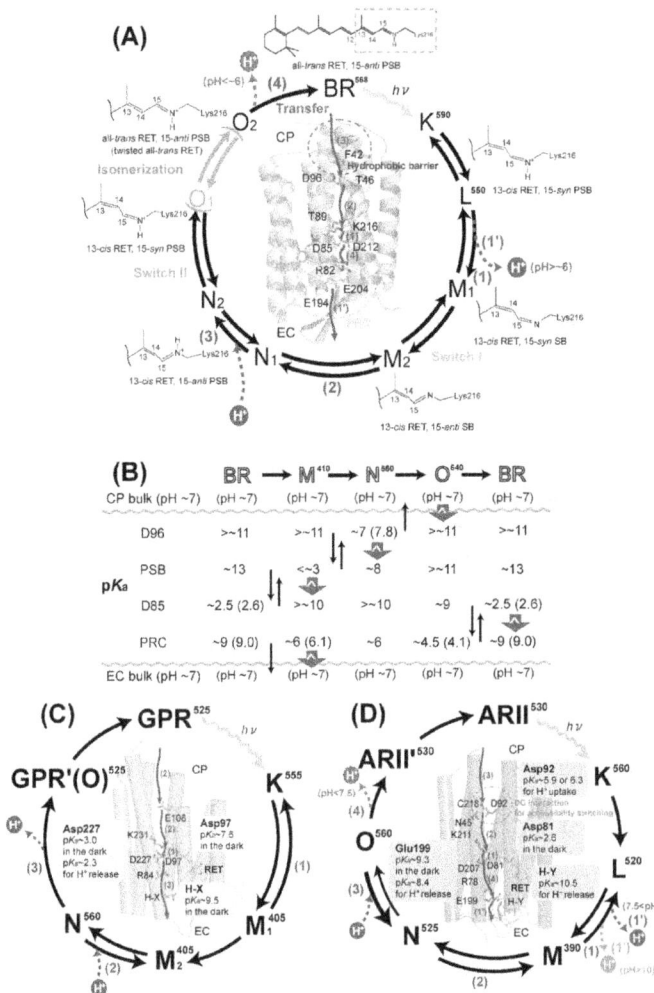

Figure 4.
Schematic representation of the photocycle and accompanying proton transfer of three types of H⁺-pumping rhodopsins. (A) Schematic diagram of the photochemistry of BR. The stepwise proton transfer reactions are depicted by thin blue arrows and overlaid on the crystal structure of BR at the dark state (PDB 1c3w). The timing of H⁺-release differs depending on pH values. The expected configuration changes of chromophore retinal (RET) and PSB in each photocycle intermediate are also depicted. (B) Summary of pK_a changes in the residues participating in the proton transfer reactions during the photocycle. A transient pK_a increase and decrease of respective residues upon each transition are shown in upward and downward thin arrows. The reverse of pK_a values between two adjacent residues leads to a unidirectional proton movement from a (protonated) residue with a lowered pK_a value to another (unprotonated) residue with an elevated pK_a value. Such proton migrations are expressed in thick blue arrows. The values in parentheses represent our previous estimated pK_a values by the SnO_2 electrode method [162]. (C) Schematic diagram of the photochemistry of a eubacterial H⁺-pump GPR. The proton transfer reactions are the case in the pH region below ca. 9.5. H-X represents an unidentified residue whose deprotonation at the initial state induces the formation of another parallel photocycle via a different M-like state (M_a) from normal M and early proton release [164]. (D) Schematic diagram of the photochemistry of a eukaryotic H⁺-pump ARII. The timing of H⁺-release is divided into three patterns depending on pH [130]: 1) an initial H⁺-release from an unknown Y-residue (H-Y) at pH > ~10 (shown in orange arrow), 2) an initial H⁺-release from $Glu199^{ARII}$ at ~ 7.5 < pH < ~10 (shown in blue arrow), and 3) a probably direct H⁺-release from $Asp81^{ARII}$ at the latter stage (O-ARII' transition) of the photocycle (shown in gray arrow). In panels C and D, the pK_a values of several crucial residues for H⁺-pumping in respective rhodopsins, which were previously estimated using the SnO_2 (or ITO) electrode method [130, 145, 164], are shown together.

In addition, studies on H^+-pumping microbial rhodopsins are required to develop novel optical cellular control methods because these types of pigments can simultaneously induce alkalization of the intracellular pH by illumination-induced outward proton transport. As is generally known, the maintenance of an appropriate cellular pH is necessary to ensure that each requisite enzyme for various biological reactions functions properly. Because the drainage of acids produced by cellular metabolism is controlled through the Na^+/H^+ antiporter or the Cl^-/HCO_3^- exchanger to maintain the cellular pH near neutral, failure in these transporting systems affects the normal function of cells. Therefore, the application of optogenetics to cells with abnormal pH values, that is, light-induced manipulation of the cells specifically expressing H^+-pumping microbial rhodopsins, may allow the restoration of the functions of these cells. As an example of intracellular pH regulation by optogenetics, Matsui et al. reported that the photoinduced intracellular pH increase in glial cells expressing aR-3 (Arch) suppressed the release of glutamate from these cells, which was triggered by glial acidosis upon brain ischemia, thereby ameliorating the effects of ischemic brain damage [167]. Moreover, the optical regulation of the function of varying organelles expressing H^+-pumping rhodopsins has recently been attempted. Rost et al. demonstrated that selective Arch expression on synaptic vesicles together with a pH-sensitive indicator and successive illumination led to vesicular acidification via Arch instead of vacuolar-type H^+-ATPases (V-ATPases), enabling neurotransmitter accumulation within synaptic vesicles driven by the proton motive force (PMF) generated through light-activated Arch [168]. In addition, Hara et al. achieved dR-2-mediated optical partial suppression of cell death induced by the inhibition of respiratory PMF generation in the mitochondria of mammalian cells [169]. More recently, a method for topological inversion of microbial rhodopsins as optogenetic tools was also developed [170]. Hence, the application of this technique together with the use of recently discovered natural inward H^+-pumping rhodopsins [171, 172] as optogenetic tools may allow the induction of both light-activated acidification and alkalization in various types of cells or organelles such as mitochondria, vesicles, and lysosomes. Hence, the combination of an outward H^+-pumping rhodopsin and the topological reversal technique described above may allow various types of optogenetics. For instance, the use of outward H^+-pumping rhodopsin might lead to the following optogenetics: in general, the pH values of lysosomes in normal cells are regulated to be approximately 5, whereas those of lysosomes in cancer cells with acquired resistance to carcinostatic agents tend to be lower [173, 174]. The efficacy of carcinostatic agents for these cancer cells is degraded because they get trapped in acidified organelles; therefore, specific expression of H^+-pumping rhodopsins in lysosomes of drug-resistant cancer cells and optical pH control (photoinduced alkalization) of these cellular organelles might lead to the restoration of the original effect of drugs. Thus, optogenetics using H^+-pumping microbial rhodopsins may lead to the establishment of new optical therapies in the future.

5. Conclusion

Proton pump-type microbial rhodopsins are not only effective neural suppressors but also optical tools for pH control of various cells or organelles that specifically incorporate these pigments, which makes them a dual optogenetic tool. Rational protein engineering based on molecular mechanisms is required to further develop these rhodopsins into more effective tools. Considering the photochemical reaction and accompanying proton transfer mechanism in various H^+-pumping rhodopsins described previously, mutations that increase their photocycle kinetics may be effective for enhancing the respective H^+-pumping abilities. To increase their H^+-pumping efficiency via their photocycles, for example, a mutation that lowers the pK_a values of proton acceptors in the unphotolyzed state, which increases the population with H^+-pumping activity, may be effective. Alternatively, alterations that lead to a reduction in the pK_a of the proton donor upon M–N transition (donor-to-SB H^+-transfer) and to an increase in its pK_a value upon N–O transition (H^+-uptake) may be efficacious for promoting CP-side proton transfer. In addition, the introduction of PRC-forming residues on the EC surface may facilitate EC proton transfer. While screening for more effective tools among such designed mutants based on their molecular mechanism, the SnO_2 (ITO) electrode method could be a simple and efficient tool for estimating the pK_a values of critical residues for proton pumps, which is an index of proton pumping effectiveness. Thus, through a series of investigations on H^+-pumping rhodopsins based on molecular mechanisms, novel optogenetic H^+-pumping rhodopsins could be developed in the near future.

Conflict of interest

The authors declare no competing financial interests.

Author details

Jun Tamogami[1*] and Takashi Kikukawa[2]

1 College of Pharmaceutical Sciences, Matsuyama University, Matsuyama, Ehime, Japan

2 Faculty of Advanced Life Science, Hokkaido University, Sapporo, Japan

*Address all correspondence to: jtamoga@g.matsuyama-u.ac.jp

IntechOpen

References

[1] Boyden ES, Zhang F, Bamberg E, Nagel G, Deisseroth K. Millisecond-timescale, genetically targeted optical control of neural activity. Nature Neuroscience. 2005;**8**:1263-1268.

[2] Nagel G, Brauner M, Liewald JF, Adeishvili N, Bamberg E, Gottschalk A. Light activation of channelrhodopsin-2 in excitable cells of *Caenorhabditis elegans* triggers rapid behavioral responses. Current Biology. 2005;**15**: 2279-2284. DOI: 10.1016/j. cub.2005.11.032

[3] Zhang F, Wang LP, Boyden ES, Deisseroth K. Channelrhodopsin-2 and optical control of excitable cells. Nature Methods. 2006;**3**:785-792. DOI: 10.1038/ NMETH936

[4] Zhang F, Wang LP, Brauner M, Liewald JF, Kay K, Watzke N, Wood PG, Bamberg E, Nagel G, Gottschalk A, Deisseroth K. Multimodal fast optical interrogation of neural circuitry. Nature. 2007;**446**:633-639. DOI: 10.1038/ nature05744

[5] Zhang F, Aravanis AM, Adamantidis A, de Lecea L, Deisseroth, K. Circuit-breakers: optical technologies for probing neural signals and systems. Nature Reviews Neuroscience. 2007;**8**: 577-581. DOI: 10.1038/nrn2192

[6] Häusser M, Smith SL. Controlling neural circuits with light. Nature. 2007; **446**:617-619. DOI: 10.1038/446617a

[7] Han X, Qian X, Bernstein JG, Zhou HH, Franzesi GT, Stern P, Bronson RT, Graybiel AM, Desimone R, Boyden ES. Millisecond-timescale optical control of neural dynamics in the nonhuman primate brain. Neuron. 2009; **62**:191-198. DOI: 10.1016/j. neuron.2009.03.011

[8] Deisseroth K, Hegemann P. The form and function of channelrhodopsin. Science. 2017;**357**:eaan5544. DOI: 10.1126/science.aan5544

[9] Gradinaru V, Zhang F, Ramakrishnan C, Mattis J, Prakash R, Diester I, Goshen I, Thompson KR, Deisseroth K. Molecular and cellular approaches for diversifying and extending optogenetics. Cell. 2010;**141**: 154-165. DOI: 10.1016/j.cell.2010.02.037

[10] Kleinlogel S, Feldbauer K, Dempski RE, Fotis H, Wood PG, Bamann C, Bamberg E. Ultra light-sensitive and fast neuronal activation with the Ca^{2+}-permeable channelrhodopsin CatCh. Nature Neuroscience. 2011;**14**:513-518. DOI: 10.1038/nn.2776

[11] Wietek J, Wiegert JS, Adeishvili N, Schneider F, Watanabe H, Tsunoda SP, Vogt A, Elstner M, Oertner TG, Hegemann P. Conversion of channelrhodopsin into a light-gated chloride channel. Science. 2014;**344**: 409-412. DOI: 10.1126/science.1249375

[12] Bedbrook CN, Yang KK, Robinson JE, Mackey ED, Gradinaru V, Arnold FH. Machine learning-guided channelrhodopsin engineering enables minimally-invasive optogenetics. Nature Methods. 2019;**16**: 1176-1184. DOI: 10.1038/s41592-019-0583-8

[13] Cho YK, Park D, Yang A, Chen F, Chuong AS, Klapoetke NC, Boyden ES. Multidimensional screening yields channelrhodopsin variants having improved photocurrent and order-of-magnitude reductions in calcium and proton currents. The Journal of Biological Chemistry. 2019;**294**:

3806-3821. DOI: 10.1074/jbc.
RA118.006996

[14] Oppermann J, Fischer P, Silapetere A, Liepe B, Rodriguez-Rozada S, Flores-Uribe J, Peter E, Keidel A, Vierock J, Kaufmann J, Broser M, Luck M, Bartl F, Hildebrandt P, Wiegert JS, Béjà O, Hegemann P, Wietek J. MerMAIDs: a family of metagenomically discovered marine anion-conducting and intensely desensitizing channelrhodopsins. Nature Communications. 2019;**10**:3315. DOI: 10.1038/s41467-019-11322-6

[15] Li X, Gutierrez DV, Hanson MG, Han J, Mark MD, Chiel H, Hegemann P, Landmesser LT, Herlitze S. Fast noninvasive activation and inhibition of neural and network activity by vertebrate rhodopsin and green algae channelrhodopsin. Proceedings of the National Academy of Sciences of the United States of America. 2005;**102**: 17816-17821. DOI: 10.1073_pnas.0509030102

[16] Levskaya A, Weiner OD, Lim WA, Voigt CA. Spatiotemporal control of cell signalling using a light-switchable protein interaction. Nature. 2009;**461**: 997-1001. DOI: 10.1038/nature08446

[17] Sierra YAB, Rost BR, Pofahl M, Fernandes AM, Kopton RA, Moser S, Holtkamp D, Masala N, Beed P, Tukker JJ, Oldani S, Bönigk W, Kohl P, Baier H, Schneider-Warme F, Hegemann P, Beck H, Seifert R, Schmitz D. Potassium channel-based optogenetic silencing. Nature Communications. 2018;**9**:4611. DOI: 10.1038/s41467-018-07038-8

[18] Bamann C, Nagel G, Bamberg E. Microbial rhodopsins in the spotlight. Current Opinion in Neurobiology. 2010; **20**:610-616. DOI: 10.1016/j. conb.2010.07.003

[19] Zhang F, Vierock J, Yizhar O, Fenno LE, Tsunoda S, Kianianmomeni A, Prigge M, Berndt A, Cushman J, Polle J, Magnuson J, Hegemann P, Deisseroth K. The microbial opsin family of optogenetic tools. Cell. 2011;**147**:1446-1457. DOI 10.1016/j.cell.2011.12.004

[20] Spudich JL, Yang CS, Jung KH, Spudich EN. Retinylidene proteins: structures and functions from archaea to humans. Annual Review of Cell and Developmental Biology. 2000;**16**: 365-392. DOI: 10.1146/annurev. cellbio.16.1.365

[21] Béjà O, Lanyi JK. Nature's toolkit for microbial rhodopsin ion pumps. Proceedings of the National Academy of Sciences of the United States of America. 2014;**111**: 6538-6539. DOI: 10.1073/ pnas.1405093111

[22] Ernst OP, Lodowski DT, Elstner M, Hegemann P, Brown LS, Kandori H. Microbial and animal rhodopsins: structures, functions, and molecular mechanisms. Chemical Reviews. 2014; **114**:126-163. DOI: 10.1021/cr4003769

[23] Grote M, Engelhard M, Hegemann P. Of ion pumps, sensors and channels — Perspectives on microbial rhodopsins between science and history. Biochimica et Biophysica Acta. 2014;**1837**:533-545. DOI: 10.1016/j.bbabio.2013.08.006

[24] Inoue K, Kato Y, Kandori H. Light-driven ion-translocating rhodopsins in marine bacteria. Trends in Microbiology. 2015;**23**:91-98. DOI: 10.1016/j. tim.2014.10.009

[25] Nagel G, Ollig D, Fuhrmann M, Kateriya S, Musti AM, Bamberg E, Hegemann P. Channelrhodopsin-1: a light-gated proton channel in green algae. Science. 2002;**296**:2395-2398. DOI: 10.1126/science.1072068

[26] Nagel G, Szellas T, Huhn W, Kateriya S, Adeishvili N, Berthold P, Ollig D, Hegemann P, Bamberg E. Channelrhodopsin-2, a directly light-gated cation-selective membrane channel. Proceedings of the National Academy of Sciences of the United States of America. 2003;**100**: 13940-13945. DOI: 10.1073_ pnas.1936192100

[27] Nagel G, Szellas T, Kateriya S, Adeishvili N, Hegemann P, Bamberg E. Channelrhodopsins: directly light-gated cation channels. Biochemical Society Transactions. 2005;**33**:863-866. DOI: 10.1042/BST0330863

[28] Govorunova EG, Sineshchekov OA, Janz R, Liu X, Spudich JL. Natural light-gated anion channels: a family of microbial rhodopsins for advanced optogenetics. Science. 2015; **349**:647-650. DOI: 10.1126/science. aaa7484

[29] Govorunova EG, Sineshchekov OA, Spudich JL. *Proteomonas sulcata* ACR1: a fast anion channelrhodopsin. Photochemistry and Photobiology. 2016;**92**:257-263. DOI: 10.1111/php.12558

[30] Govorunovaa EG, Sineshchekova OA, Lia H, Wanga Y, Brownb LS, Spudicha JL. RubyACRs, nonalgal anion channelrhodopsins with highly red-shifted absorption. Proceedings of the National Academy of Sciences of the United States of America. 2020;**117**:22833-22840. DOI: 10.1073/ pnas.2005981117

[31] Govorunova EG, Sineshchekov OA, Hemmati R, Janz R, Morelle O, Melkonian M, Wong GKS, Spudich JL. Extending the time domain of neuronal silencing with cryptophyte anion channelrhodopsins. eNeuro. 2018;**5**: ENEURO.0174-18.2018. DOI: 10.1523/ ENEURO.0174-18.2018

[32] Haupts U, Tittor J, Oesterhelt D. Closing in on bacterorhodopsin: progress in understanding the molecule. Annual Review of Biophysics and Biomolecular Structure. 1999;**28**: 367-399. DOI: 10.1146/annurev. biophys.28.1.367

[33] Lanyi JK. Bacteriorhodopsin. Annual Review of Physiology. 2004;**66**:665-688. DOI: 10.1146/annurev. physiol.66.032102.150049

[34] Inoue K, Ono H, Abe-Yoshizumi R, Yoshizawa S, Ito H, Kogure K, Kandori H. A light-driven sodium ion pump in marine bacteria. Nature Communications. 2013;**4**:1678. DOI: 10.1038/ncomms2689

[35] Mukohata Y, Ihara K, Tamura T, Sugiyama Y. Halobacterial rhodopsins. The Journal of Biochemistry. 1999;**125**: 649-657. DOI: 10.1093/oxfordjournals. jbchem.a022332

[36] Váró G. Analogies between halorhodopsin and bacteriorhodopsin. Biochimica et Biophysica Acta. 2000; **1460**:220-229. DOI: 10.1016/S0005-2728 (00)00141-9

[37] Essen LO. Halorhodopsin: light-driven ion pumping made simple? Current Opinion in Structural Biology. 2002;**12**:516-522. DOI: 10.1016/ S0959-440X(02)00356-1

[38] Engelhard C, Chizhov I, Siebert F, Engelhard M. Microbial halorhodopsins: light-driven chloride pumps. Chemical Reviews. 2018;**118**:10629-10645. DOI: 10.1021/acs.chemrev.7b00715

[39] Chow BY, Han X, Dobry AS, Qian X, Chuong AS, Li M, Henninger MA, Belfort GM, Lin Y, Monahan PE, Boyden ES. High-performance genetically targetable optical neural silencing by light-driven proton pumps.

Nature. 2010;**463**:98-102. DOI: 10.1038/nature08652

[40] Grimm C, Silapetere A, Vogt A, Bernal Sierra YA, Hegemann P. Electrical properties, substrate specificity and optogenetic potential of the engineered light-driven sodium pump eKR2. Scientific Reports. 2018;**8**:9316. DOI: 10.1038/s41598-018-27690-w

[41] Finkel OM, Béjà O, Belkin S. Global abundance of microbial rhodopsins. The ISME Journal. 2013;7:448-451. DOI: 10.1038/ismej.2012.112

[42] Oesterhelt D, Stoeckenius W. Rhodopsin-like Protein from the Purple Membrane of *Halobacterium halobium*. Nature New Biology. 1971;**233**:149-152. DOI: 10.1038/newbio233149a0

[43] Subramaniam S. The structure of bacteriorhodopsin: an emerging consensus. Current Opinion in Structural Biology. 1999;**9**:462-468. DOI: 10.1016/S0959-440X(99)80065-7

[44] Heberle J. Proton transfer reactions across bacteriorhodopsin and along the membrane. Biochimica et Biophysica Acta. 2000;**1458**:135-147. DOI: 10.1016/s0005-2728(00)00064-5

[45] Balashov SP. Protonation reactions and their coupling in bacteriorhodopsin. Biochimica et Biophysica Acta. 2000;**1460**:75-94. DOI: 10.1016/s0005-2728(00)00131-6

[46] Subramaniam S, Henderson R. Crystallographic analysis of protein conformational changes in the bacteriorhodopsin photocycle. Biochimica et Biophysica Acta. 2000;**1460**:157-165. DOI: 10.1016/s0005-2728(00)00136-5

[47] Kandori H. Role of internal water molecules in bacteriorhodopsin. Biochimica et Biophysica Acta. 2000;**1460**:177-191. DOI: 10.1016/s0005-2728(00)00138-9

[48] Wickstrand C, Nogly P, Nango E, Iwata S, Standfuss J, Neutze R. Bacteriorhodopsin: structural insights revealed using X-ray lasers and synchrotron radiation. Annual Review of Biochemistry. 2019;**88**:59-83. DOI: 10.1146/annurev-biochem-013118-111327

[49] Mukohata Y, Sugiyama Y, Ihara K, Yoshida M. An Australian halobacterium contains a novel proton pump retinal protein: archaerhodopsin. Biochemical and Biophysical Research Communications. 1988;**151**:1339-1345. DOI: 10.1016/S0006-291X(88)80509-6

[50] Ihara K, Umemura T, Katagiri I, Kitajima-Ihara T, Sugiyama Y, Kimura Y, Mukohata Y. Evolution of the archaeal rhodopsins: evolution rate changes by gene duplication and functional differentiation. Journal of Molecular Biology. 1999;**285**:163-174. DOI: 10.1006/jmbi.1998.2286

[51] Li Q, Sun Q, Zhao W, Wang H, Xu D. Newly isolated archaerhodopsin from a strain of Chinese halobacteria and its proton pumping behavior. Biochimica et Biophysica Acta. 2000;**1466**:260-266. DOI: 10.1016/S0005-2736(00)00188-7

[52] Chaoluomeng, Dai G, Kikukawa T, Ihara K, Iwasa T (2015). Microbial rhodopsins of *Halorubrum* species isolated from Ejinoor salt lake in Inner Mongolia of China. Photochemical & Photobiological Sciences. 2015;**14**:1974-82. DOI: 10.1039/c5pp00161g

[53] Geng X, Dai G, Chao L, Wen D, Kikukawa T, Iwasa T. Two consecutive polar amino acids at the end of helix E are important for fast turnover of the archaerhodopsin photocycle. Photochemistry and Photobiology. 2019;**95**:980-989. DOI: 10.1111/php.13072

[54] Tateno M, Ihara K, Mukohata Y. The novel ion pump rhodopsins from *Haloarcula* form a family independent from both the bacteriorhodopsin and archaerhodopsin families/tribes. Archives of Biochemistry and Biophysics. 1994;**315**:127-132. DOI: 10.1006/abbi.1994.1480

[55] Sugiyama Y, Yamada N, Mukohata Y. The light-driven proton pump, cruxrhodopsin-2 in *Haloarcula* sp. arg-2 (bR$^+$, hR$^-$), and its coupled ATP formation. Biochimica et Biophysica Acta. 1994;**1188**:287-292. DOI: 10.1016/0005-2728(94)90047-7

[56] Kitajima T, Hirayama J, Ihara K, Sugiyama Y, Kamo N. Novel bacterial rhodopsins from *Haloarcula vallismortis*. Biochemical and Biophysical Research Communications. 1996;**345**:341-345. DOI: 10.1006/bbrc.1996.0407

[57] Kamekura M, Seno Y, Tomioka H. Detection and expression of a gene encoding a new bacteriorhodopsin from an extreme halophile strain HT (JCM 9743) which does not possess bacteriorhodopsin activity. Extremophiles. 1998;**2**:33-39. DOI: 10.1007/s007920050040

[58] Brown LS, Jung KH. Bacteriorhodopsin-like proteins of eubacteria and fungi: the extent of conservation of the haloarchaeal proton-pumping mechanism. Photochemical & Photobiological Sciences. 2006;**5**: 538-546. DOI: 10.1039/b514537f

[59] Brown LS. Eubacterial rhodopsins — Unique photosensors and diverse ion pumps. Biochimica et Biophysica Acta. 2014;**1837**:553-561. DOI: 10.1016/j.bbabio.2013.05.006

[60] Fuhrman JA, Schwalbach MS, Stingl U. Proteorhodopsins: an array of physiological roles? Nature Reviews Mic

robiology. 2008;**6**:488-494. DOI: 10.1038/nrmicro1893

[61] Bamann C, Bamberg E, Wachtveitl J, Glaubitz C. Proteorhodopsin. Biochimica et Biophysica Acta. 2014;**1837**:614-625. DOI: 10.1016/j.bbabio.2013.09.010

[62] Béjà O, Aravind L, Koonin EV, Suzuki MT, Hadd A, Nguyen LP, Jovanovich SB, Gates CM, Feldman RA, Spudich JL, Spudich EN, DeLong EF. Bacterial rhodopsin: evidence for a new type of phototrophy in the Sea. Science. 2000;**289**:1902-1906. DOI: 10.1126/science.289.5486.1902

[63] Giovannoni SJ, Bibbs L, Cho JC, Stapels MD, Desiderio R, Vergin KL, Rappé MS, Laney S, Wilhelm LJ, Tripp HJ, Mathur EJ, Barofsky DF. Proteorhodopsin in the ubiquitous marine bacterium SAR11. Nature. 2005; **438**:82-85. DOI: 10.1038/nature04032

[64] Courties A, Riedel T, Jarek M, Intertaglia L, Lebaron P, Suzuki MT. Genome sequence of strain MOLA814, a proteorhodopsin-containing representative of the *Betaproteobacteria* common in the ocean. Genome Announcements. 2013;**1**:e01062-e01013. DOI: 10.1128/genomeA.01062-13

[65] Gómez-Consarnau L, González JM, Coll-Lladó M, Gourdon P, Pascher T, Neutze R, Pedrós-Alió C, Pinhassi J. Light stimulates growth of proteorhodopsin-containing marine Flavobacteria. Nature. 2007;**445**: 210-213. DOI: 10.1038/nature05381

[66] Zhao M, Chen F, Jiao N. Genetic diversity and abundance of Flavobacterial proteorhodopsin in China Seas. Applied and Environmental Microbiology. 2009; **75**:529-533. DOI: 10.1128/AEM.01114-08

[67] Béjà O, Spudich EN, Spudich JL, Leclerc M, DeLong EF. Proteorhodopsin

phototrophy in the ocean. Nature. 2001; **411**:786-789. DOI: 10.1038/35081051

[68] Sabehi G, Massana R, Bielawski JP, Rosenberg M, Delong EF, Béjà O. Novel proteorhodopsin variants from the Mediterranean and Red Seas. Environmental Microbiology. 2003;**5**: 842-849. DOI: 10.1046/ j.1462-2920.2003.00493.x

[69] de la Torre JR, Christianson LM, Béjà O, Suzuki MT, Karl DM, Heidelberg J, DeLong EF. Proteorhodopsin genes are distributed among divergent marine bacterial taxa. Proceedings of the National Academy of Sciences of the United States of America. 2003;**100**:12830-12835. DOI: 10.1073_ pnas.2133554100

[70] Venter JC, Remington K, Heidelberg JF, Halpern AL, Rusch D, Eisen JA, Wu D, Paulsen I, Nelson KE, Nelson W, Fouts DE, Levy S, Knap AH, Lomas MW, Nealson K, White O, Peterson J, Hoffman J, Parsons R, Baden-Tillson H, Pfannkoch C, Rogers YH, Smith HO. Environmental genome shotgun sequencing of the Sargasso Sea. Science. 2004;**304**:66-74. DOI: 10.1126/ science.1093857

[71] Rusch DB, Halpern AL, Sutton G, Heidelberg KB, Williamson S, Yooseph S, Wu D, Eisen JA, Hoffman JM, Remington K, Beeson K, Tran B, Smith H, Baden-Tillson H, Stewart C, Thorpe J, Freeman J, Andrews-Pfannkoch C, Venter JE, Li K, Kravitz S, Heidelberg JF, Utterback T, Rogers YH, Falcón LI, Souza V, Bonilla-Rosso G, Eguiarte LE, Karl DM, Sathyendranath S, Platt T, Bermingham E, Gallardo V, Tamayo-Castillo G, Ferrari MR, Strausberg RL, Nealson K, Friedman R, Frazier M, Venter JC. The sorcerer II global ocean sampling expedition: northwest Atlantic through eastern tropical Pacific. PLoS

Biology. 2007;**5**:e77. DOI: 10.1371/ journal.pbio.0050077

[72] Yoshizawa S, Kawanabe A, Ito H, Kandori H, Kogure K. Diversity and functional analysis of proteorhodopsin in marine *Flavobacteria*. Environmental Microbiology. 2012;**14**:1240-1248. DOI: 10.1111/j.1462-2920.2012.02702.x

[73] Man D, Wang W, Sabehi G, Aravind L, Post AF, Massana R, Spudich EN, Spudich JL, Béjà O. Diversification and spectral tuning in marine proteorhodopsins. The EMBO Journal. 2003;**22**:1725-1731. DOI: 10.1093/emboj/cdg183

[74] Bielawski JP, Dunn KA, Sabehi G, Béjà O. Darwinian adaptation of proteorhodopsin to different light intensities in the marine environment. Proceedings of the National Academy of Sciences of the United States of America. 2004;**101**:14824-14829. DOI: 10.1073_ pnas.0403999101

[75] Sabehi G, Kirkup BC, Rozenberg M, Stambler N, Polz MF, Béjà O. Adaptation and spectral tuning in divergent marine proteorhodopsins from the eastern Mediterranean and the Sargasso Seas. The ISME Journal. 2007;**1**: 48-55. DOI: 10.1038/ismej.2007.10

[76] Sharma AK, Zhaxybayeva O, Papke RT, Doolittle WF. Actinorhodopsins: proteorhodopsin-like gene sequences found predominantly in non-marine environments. Environmental Microbiology. 2008;**10**:1039-1056. DOI: 10.1111/j.1462-2920.2007.01525.x

[77] Bohorquez LC, Ruiz-Perez CA, Zambrano MM. Proteorhodopsin-like genes present in thermoacidophilic high-mountain microbial communities. Applied and Environmental Microbiolog y. 2012;**78**:7813-7817. DOI: 10.1128/ AEM.01683-12

[78] Murugapiran SK, Huntemann M, Wei CL, Han J, Detter JC, Han CS, Erkkila TH, Teshima H, Chen A, Kyrpides N, Mavrommatis K, Markowitz V, Szeto E, Ivanova N, Pagani I, Lam J, McDonald AI, Dodsworth JA, Pati A, Goodwin L, Peters L, Pitluck S, Woyke T, Hedlund BP. Whole genome sequencing of *Thermus oshimai* JL-2 and *Thermus thermophilus* JL-18, incomplete denitrifiers from the United States Great Basin. Genome Announcements. 2013;**1**: e00106-e00112. DOI: 10.1128/genomeA.00106-12

[79] Petrovskaya LE, Lukashev EP, Chupin VV, Sychev SV, Lyukmanova EN, Kryukova EA, Ziganshin RH, Spirina EV, Rivkina EM, Khatypov RA, Erokhina LG, Gilichinsky DA, Shuvalov VA, Kirpichnikov MP. Predicted bacteriorhodopsin from *Exiguobacterium sibiricum* is a functional proton pump. FEBS Letters. 2010;**584**:4193-4196. DOI: 10.1016/j.febslet.2010.09.005

[80] Balashov SP, Imasheva ES, Boichenko VA, Antón J, Wang JM, Lanyi JK. Xanthorhodopsin: a proton pump with a light-harvesting carotenoid antenna. Science. 2005;**309**:2061-2064. DOI: 10.1126/science.1118046

[81] Lanyi JK, Balashov SP. Xanthorhodopsin: a bacteriorhodopsin-like proton pump with a carotenoid antenna. Biochimica et Biophysica Acta. 2008;**1777**:684-688. DOI: 10.1016/j.bbabio.2008.05.005

[82] Imasheva ES, Balashov SP, Choi AR, Jung KH, Lanyi JK. Reconstitution of *Gloeobacter violaceus* rhodopsin with a light-harvesting carotenoid antenna. Biochemistry. 2009;**48**:10948-10955. DOI: 10.1021/bi901552x

[83] Miranda MRM, Choi AR, Shi L, Bezerra Jr. AG, Jung KH, Brown LS. The

photocycle and proton translocation pathway in a cyanobacterial ion-pumping rhodopsin. Biophysical Journal. 2009;**96**:1471-1481. DOI: 10.1016/j.bpj.2008.11.026

[84] Frigaard NU, Martinez A, Mincer TJ, DeLong EF. Proteorhodopsin lateral gene transfer between marine planktonic Bacteria and Archaea. Nature. 2006;**439**:847-850. DOI: 10.1038/nature04435

[85] Iverson V, Morris RM, Frazar CD, Berthiaume CT, Morales RL, Armbrust EV. Untangling genomes from metagenomes: revealing an uncultured class of marine euryarchaeota. Science. 2012;**335**:587-590. DOI: 10.1126/science.1212665.

[86] Slamovits CH, Okamoto N, Burri L, James ER, Keeling PJ. A bacterial proteorhodopsin proton pump in marine eukaryotes. Nature Communications. 2011;**2**:183. DOI: 10.1038/ncomms1188

[87] Bieszke JA, Braun EL, Bean LE, Kang S, Natvig DO, Borkovich KA. The *nop-1* gene of *Neurospora crassa* encodes a seven transmembrane helix retinal-binding protein homologous to archaeal rhodopsins. Proceedings of the National Academy of Sciences of the United States of America. 1999;**96**:8034-8039. DOI: 10.1073/pnas.96.14.8034

[88] Bieszke JA, Spudich EN, Scott KL, Borkovich KA, Spudich JL. A eukaryotic protein, NOP-1, binds retinal to form an archaeal rhodopsin-like photochemically reactive pigment. Biochemistry. 1999;**38**: 14138-14145. DOI: 10.1021/bi9916170

[89] Bieszke JA, Li L, Borkovich KA. The fungal opsin gene nop-1 is negatively-regulated by a component of the blue light sensing pathway and influences conidiation-specific gene expression in *Neurospora crassa*. Current Genetics.

2007;**52**:149-157. DOI: 10.1007/s00294-007-0148-8

[90] Estrada AF, Avalos J. Regulation and targeted mutation of opsA, coding for the NOP-1 opsin orthologue in *Fusarium fujikuroi*. Journal of Molecular Biology. 2009;**387**:59-73. DOI: 10.1016/j.jmb.2009.01.057

[91] Waschuk SA, Bezerra AG, Shi L, Brown LS. *Leptosphaeria* rhodopsin: bacteriorhodopsin-like proton pump from a eukaryote. Proceedings of the National Academy of Sciences of the United States of America. 2005;**102**:6879-6883. DOI: 10.1073/pnas.0409659102

[92] Fan Y, Solomon P, Oliver RP, Brown LS. Photochemical characterization of a novel fungal rhodopsin from *Phaeosphaeria nodorum*. Biochimica et Biophysica Acta. 2011; **1807**:1457-1466. DOI: 10.1016/j.bbabio.2011.07.005

[93] Tsunoda SP, Ewers D, Gazzarrini S, Moroni A, Gradmann D, Hegemann P. H⁺-pumping rhodopsin from the marine alga *Acetabularia*. Biophysical Journal. 2006;**91**:1471-1479. DOI: 10.1529/biophysj.106.086421

[94] Henry IM, Wilkinson MD, Hernandez JM, Schwarz-Sommer Z, Grotewold E. Mandoli DF. Comparison of ESTs from juvenile and adult phases of the giant unicellular green alga *Acetabularia acetabulum*. BMC Plant Biology. 2004;**4**:3. DOI: 10.1186/1471-2229-4-3

[95] Lee SS, Choi AR, Kim SY, Kang HW, Jung KH, Lee JH. *Acetabularia* rhodopsin I is a light-stimulated proton pump. Journal of Nanoscience and Nanotechnology. 2011;**11**:4596-4600. DOI:10.1166/jnn.2011.3650

[96] Wada T, Shimono K, Kikukawa T, Hato M, Shinya N, Kim SY,

Kimura-Someya T, Shirouzu M, Tamogami J, Miyauchi S, Jung KH, Kamo N, Yokoyama S. Crystal structure of the eukaryotic light-driven proton-pumping rhodopsin, *Acetabularia* rhodopsin II, from marine alga. Journal of Molecular Biology. 2011;**411**:986-998. DOI: 10.1016/j.jmb.2011.06.028

[97] Mogi T, Stern LJ, Marti T, Chao BH, Khorana HG. Aspartic acid substitutions affect proton translocation by bacteriorhodopsin. Proceedings of the National Academy of Sciences of the United States of America. 1988;**85**: 4148-4152. DOI: 10.1073/pnas.85.12.4148

[98] Garczarek F, Brown LS, Lanyi JK, Gerwert K. Proton binding within a membrane protein by a protonated water cluster. Proceedings of the National Academy of Sciences of the United States of America. 2005;**102**: 3633-3638. DOI: 10.1073/pnas.0500421102

[99] Garczarek F, Gerwert K. Functional waters in intraprotein proton transfer monitored by FTIR difference spectroscopy. Nature. 2006;**439**:109-112. DOI: 10.1038/nature04231

[100] Zimányi L, Váró G, Chang M, Ni B, Needleman R, Lanyi JK. Pathways of proton release in the bacteriorhodopsin photocycle. Biochemistry. 1992;**31**: 8535-8543. DOI: 10.1021/bi00151a022

[101] Balashov SP, Lu M, Imasheva ES, Govindjee R, Ebrey TG, Othersen III B, Chen Y, Crouch RK, Menick DR. The proton release group of bacteriorhodopsin controls the rate of the final step of its photocycle at low pH. Biochemistry. 1999;**38**:2026-2039. DOI: 10.1021/bi981926a

[102] Russell TS, Coleman M, Rath P, Nilsson A, Rothschild KJ. Threonine-89

participates in the active site of bacteriorhodopsin: evidence for a role in color regulation and Schiff base proton transfer. Biochemistry. 1997;**36**: 7490-7497. DOI: 10.1021/bi9702871

[103] Kandori H, Yamazaki Y, Shichida Y, Raap J, Lugtenburg J, Belenky M, Herzfeld J. Tight Asp-85–Thr-89 association during the pump switch of bacteriorhodopsin. Proceedings of the National Academy of Sciences of the United States of America. 2001;**98**:1571-1576. DOI: 10.1073/pnas.98.4.1571

[104] Luecke H. Atomic resolution structures of bacteriorhodopsin photocycle intermediates: the role of discrete water molecules in the function of this light-driven ion pump. Biochimica et Biophysica Acta. 2000; **1460**:133-156. DOI: 10.1016/S0005-2728 (00)00135-3

[105] Dioumaev AK, Brown LS, Shih J, Spudich EN, Spudich JL, Lanyi JK. Proton transfers in the photochemical reaction cycle of proteorhodopsin. Biochemistry. 2002;**41**:5348-5358. DOI: 10.1021/bi025563x

[106] Friedrich T, Geibel S, Kalmbach R, Chizhov I, Ataka K, Heberle J, Engelhard M, Bamberg E. Proteorhodopsin is a light-driven proton pump with variable vectoriality. Journal of Molecular Biology. 2002;**321**:821-838. DOI: 10.1016/S0022283602006964

[107] Krebs RA, Dunmire D, Partha R, Braiman MS. Resonance Raman characterization of proteorhodopsin's chromophore environment. The Journal of Physical Chemistry B. 2003;107: 7877-7883. DOI: 10.1021/jp034574c

[108] Váró G, Brown LS, Lakatos M, Lanyi JK. Characterization of the photochemical reaction cycle of

proteorhodopsin. Biophysical Journal. 2003;**84**:1202-1207. DOI: 10.1016/ S0006-3495(03)74934-0

[109] Lakatos M, Lanyi JK, Szakács J, Váró G. The photochemical reaction cycle of proteorhodopsin at low pH. Biophysical Journal. 2003;**84**:3252-3256. DOI: 10.1016/S0006-3495(03)70049-6

[110] Bergo V, Amsden JJ, Spudich EN, Spudich JL, Rothschild KJ. Structural changes in the photoactive site of proteorhodopsin during the primary photoreaction. Biochemistry. 2004;**43**: 9075-9083. DOI: 10.1021/bi0361968

[111] Ludmann K, Gergely C, Váró G. Kinetic and thermodynamic study of the bacteriorhodopsin photocycle over a wide pH range. Biophysical Journal. 1998;**75**: 3110-3119. DOI: 10.1016/ S0006-3495(98)77752-5

[112] Fujisawa T, Abe M, Tamogami J, Kikukawa T, Kamo N, Unno M. Low-temperature Raman spectroscopy reveals small chromophore distortion in primary photointermediate of proteorhodopsin. FEBS Letters. 2018;**592**:3054-3061. DOI: 10.1002/1873-3468.13219

[113] Wang WW, Sineshchekov OA, Spudich EN, Spudich JL. Spectroscopic and photochemical characterization of a deep ocean proteorhodopsin. The Journal of Biological Chemistry. 2003; **278**:33985-33991. DOI: 10.1074/jbc. M305716200

[114] Balashov SP, Petrovskaya LE, Imasheva ES, Lukashev EP, Dioumaev AK, Wang JM, Sychev SV, Dolgikh DA, Rubin AB, Kirpichnikov MP, Lanyi JK. Breaking the carboxyl rule: Lysine 96 facilitates reprotonation of the Schiff base in the photocycle of a retinal protein from *Exiguobacterium sibiricum*. The Journal of Biological Chemistry. 2013;**288**:

21254-21265 DOI: 10.1074/jbc. M113.465138

[115] Petrovskaya LE, Balashov SP, Lukashev EP, Imasheva ES, Gushchin IY, Dioumaev AK, Rubin AB, Dolgikh DA, Gordeliy VI, Lanyi JK, Kirpichnikov MP. ESR - a retinal protein with unusual properties from *Exiguobacterium sibiricum*. Biochemistry (Moscow). 2015;**80**:688-700. DOI: 10.1134/S000629791506005X

[116] Nakamura S, Kikukawa T, Tamogami J, Kamiya M, Aizawa T, Hahn MW, Ihara K, Kamo N, Demura M. Photochemical characterization of actinorhodopsin and its functional existence in the natural host. Biochimica et Biophysica Acta. 2016;**1857**:1900-1908. DOI: 10.1016/j. bbabio.2016.09.006

[117] Klyszejko AL, Shastri S, Mari SA, Grubmüller H, Muller DJ, Glaubitz C. Folding and assembly of proteorhodopsin. Journal of Molecular Biology. 2008;**376**: 35-41. DOI: 10.1016/j.jmb.2007.11.030

[118] Luecke H, Schobert B, Stagno J, Imasheva ES, Wang JM, Balashov SP, Lanyi JK. Crystallographic structure of xanthorhodopsin, the light-driven proton pump with a dual chromophore. Proceedings of the National Academy of Sciences of the United States of America. 2008;**105**:16561-16565. DOI: 10.1073_ pnas.0807162105

[119] Pfleger N, Wörner AC, Yang J, Shastri S, Hellmich UA, Aslimovska L, Maier MSM, Glaubitz C. Solid-state NMR and functional studies on proteorhodopsin. Biochimica et Biophysica Acta. 2009;**1787**:697-705. DOI: 10.1016/j.bbabio.2009.02.022

[120] Reckel S, Gottstein D, Stehle J, Lchr F, Verhoefen MK, Takeda M, Silvers R, Kainosho M, Glaubitz C,

Wachtveitl J, Bernhard F, Schwalbe H, Güntert P, Dötsch V. Solution NMR structure of proteorhodopsin. Angewandte Chemie International Edition. 2011;**50**:1-6. DOI: 10.1002/ anie.201105648

[121] Ran T, Ozorowski G, Gao Y, Sineshchekov OA, Wang W, Spudich JL, Luecke H. Cross-protomer interaction with the photoactive site in oligomeric proteorhodopsin complexes. Acta Crystallographica Section D: Structural Biology. 2013;**D69**:1965-1980. DOI: 10.1107/S0907444913017575

[122] Gushchin I, Chervakov P, Kuzmichev P, Popov AN, Round E, Borshchevskiy V, Ishchenko A, Petrovskaya L, Chupin V, Dolgikh DA, Arseniev AS, Kirpichnikov M, Gordeliy V. Structural insights into the proton pumping by unusual proteorhodopsin from nonmarine bacteria. Proceedings of the National Academy of Sciences of the United States of America. 2013;**110**:12631-12636. DOI: 10.1073/pnas.1221629110

[123] Morizumi T, Ou WL, Eps NV, Inoue K, Kandori H, Brown LS, Ernst OP. X-ray crystallographic structure and oligomerization of *Gloeobacter* rhodopsin. Scientific Reports. 2019;**9**:11283. DOI: 10.1038/ s41598-019-47445-5

[124] Brown LS. Fungal rhodopsins and opsin-related proteins: eukaryotic homologues of bacteriorhodopsin with unknown functions. Photochemical & Photobiological Sciences. 2004;**3**: 555-565. DOI: 10.1039/ b315527g

[125] Sumii M, Furutani Y, Waschuk SA, Brown LS, Kandori H. Strongly hydrogen-bonded water molecule present near the retinal chromophore of *Leptosphaeria* rhodopsin, the bacteriorhodopsin-like proton pump

from a eukaryote. Biochemistry. 2005; **44**:15159-15166. DOI: 10.1021/bi0513498

[126] Furutani Y, Sumii M, Fan Y, Shi L, Waschuk SA, Brown LS, Kandori H. Conformational coupling between the cytoplasmic carboxylic acid and the retinal in a fungal light-driven proton pump. Biochemistry. 2006;**45**: 15349-15358. DOI: 10.1021/bi0618641

[127] Fan Y, Shi L, Brown LS. Structural basis of diversification of fungal retinal proteins probed by site-directed mutagenesis of *Leptosphaeria* rhodopsin. FEBS Letters. 2007;**581**:2557-2561. DOI: 10.1016/j.febslet.2007.05.001

[128] Ito H, Sumii M, Kawanabe A, Fan Y, Furutani Y, Brown LS, Kandori H. Comparative FTIR study of a new fungal rhodopsin. The Journal of Physical Chemistry B. 2012;**116**:11881-11889. DOI: 10.1021/jp306993a

[129] Shimono K, Goto M, Kikukawa T, Miyauchi S, Shirouzu M, Kamo N, Yokoyama S. Production of functional bacteriorhodopsin by an *Escherichia coli* cell-free protein synthesis system supplemented with steroid detergent and lipid. Protein Science. 2009;**18**: 2160-2171. DOI: 10.1002/pro.230

[130] Kikukawa T, Shimono K, Tamogami J, Miyauchi S, Kim SY, Kimura-Someya T, Shirouzu M, Jung KH, Yokoyama S, Kamo N. Photochemistry of *Acetabularia* rhodopsin II from a marine plant, *Acetabularia acetabulum*. Biochemistry. 2011;**50**:8888-8898. DOI: 10.1021/bi2009932

[131] Furuse M, Tamogami J, Hosaka T, Kikukawa T, Shinya N, Hato M, Ohsawa N, Kim SY, Jung KH, Demura M, Miyauchi S, Kamo N, Shimono K, Kimura-Someya T, Yokoyama S, Shirouzu M. Structural

basis for the slow photocycle and late proton release in *Acetabularia* rhodopsin I from the marine plant *Acetabularia acetabulum*. Acta Crystallographica Section D: Structural Biology. 2015;D**71**: 2203-2216. DOI: 10.1107/ S1399004715015722

[132] Tamogami J, Kikukawa T, Nara1 T, Demura M, Kimura-Someya T, Shirouzu M, Yokoyama S, Miyauchi S, Shimono K, Kamo N. Existence of two O-like intermediates in the photocycle of *Acetabularia* rhodopsin II, a light-driven proton pump from a marine alga. Biophysics and Physicobiology. 2017;**14**: 49-55. DOI: 10.2142/biophysico.14.0_49

[133] Tamogami J, Kikukawa T, Ohkawa K, Ohsawa N, Nara T, Demura M, Miyauchi S, Kimura-Someya T, Shirouzu M, Yokoyama S, Shimono K, Kamo N. Interhelical interactions between D92 and C218 in the cytoplasmic domain regulate proton uptake upon N-decay in the proton transport of *Acetabularia* rhodopsin II. Journal of Photochemistry & Photobiology, B: Biology. 2018;**183**: 35-45. DOI: 10.1016/j. jphotobiol.2018.04.012

[134] Balashov SP, lmasheva ES, Govindjee R, Ebrey TG. Titration of aspartate-85 in bacteriorhodopsin: What it says about chromophore isomerization and proton release. Biophysical Journal. 1996;**70**:473-481. DOI: 10.1016/ S0006-3495(96)79591-7

[135] Imasheva ES, Balashov SP, Jennifer M. Wang JM, Lanyi JK. pH-dependent transitions in xanthorhodopsin. Photochemistry and Photobiology. 2006;**82**:1406-1413. DOI: 10.1562/2006-01-15-RA-776

[136] Tsukamoto T, Kikukawa T, Kurata T, Jung KH, Kamo N, Demura M. Salt bridge in the conserved His–Asp

cluster in *Gloeobacter* rhodopsin contributes to trimer formation. FEBS Letters. 2013;**587**:322-327. DOI: 10.1016/j.febslet.2012.12.022

[137] Balashov SP, Petrovskaya LE, Lukashev EP, Imasheva ES, Dioumaev AK, Wang JM, Sychev SV, Dolgikh DA, Rubin AB, Kirpichnikov MP, Lanyi JK. Aspartate-histidine interaction in the retinal Schiff base counterion of the light-driven proton pump of *Exiguobacterium sibiricum*. Biochemistry. 2012;**51**: 5748-5762. DOI: 10.1021/bi300409m Biochemistry

[138] Partha R, Krebs R, Caterino TL, Braiman MS. Weakened coupling of conserved arginine to the proteorhodopsin chromophore and its counterion implies structural differences from bacteriorhodopsin. Biochimica et Biophysica Acta. 2005;**1708**:6-12. DOI: 10.1016/j.bbabio.2004.12.009

[139] Hempelmann F, Hölper S, Verhoefen MK, Woerner AC, Köhler T, Fiedler SA, Pfleger N, Wachtveitl J, Glaubitz C. His75-Asp97 cluster in green proteorhodopsin. Journal of the American Chemical Society. 2011;**133**:4645-4654. DOI: 10.1021/ja111116a

[140] Lazarova T, Sanz C, Querol E, Padrós E. Fourier transform infrared evidence for early deprotonation of Asp85 at alkaline pH in the photocycle of bacteriorhodopsin mutants containing E194Q. Biophysical Journal. 2000;**78**: 2022-2030. DOI: 10.1016/S0006-3495 (00)76749-X

[141] Imasheva ES, Balashov SP, Wang JM, Dioumaev AK, Lanyi JK. Selectivity of retinal photoisomerization in proteorhodopsin is controlled by aspartic acid 227. Biochemistry. 2004;**43**: 1648-1655. DOI: 10.1021/bi0355894

[142] Xiao Y, Partha R, Krebs R, Braiman M. Time-resolved FTIR spectroscopy of the photointermediates involved in fast transient H^+ release by proteorhodopsin. The Journal of Physical Chemistry B. 2005;**109**:634-641. DOI: 10.1021/jp046314g

[143] Bergo VB, Sineshchekov OA, Kralj JM, Partha R, Spudich EN, Rothschild KJ, Spudich JL. His-75 in proteorhodopsin, a novel component in light-driven proton translocation by primary pumps. The Journal of Biological Chemistry. 2009;**284**:2836–2843. DOI: 10.1074/jbc.M803792200

[144] Ikeda D, Furutani Y, Kandori H. FTIR study of the retinal Schiff base and internal water molecules of proteorhodopsin. Biochemistry. 2007; **46**:5365-5373. DOI: 10.1021/bi700143g

[145] Tamogami T, Kikukawa T, Nara T, Shimono K, Demura M, Kamo N. Photoinduced proton release in proteorhodopsin at low pH: the possibility of a decrease in the pK_a of Asp227. Biochemistry. 2012;**51**: 9290-9301. DOI: 10.1021/bi300940p

[146] Bondar AN, Elstner M, Suhai S, Smith JC, Fischer S. Mechanism of primary proton transfer in bacteriorhodopsin. Structure. 2004;**12**: 1281-1288. DOI: 10.1016/j.str.2004.04.016

[147] Brown LS, Lanyi JK. Determination of the transiently lowered pK_a of the retinal Schiff base during the photocycle of bacteriorhodopsin. Proceedings of the National Academy of Sciences of the United States of America. 1996;**93**: 1731-1734. DOI: 10.1073/pnas.93.4.1731

[148] Szaraz S, Oesterhelt D, Ormos P. pH-induced structural changes in bacteriorhodopsin studied by Fourier transform infrared spectroscopy.

Biophysical Journal. 1994;**67**: 1706-1712. DOI: 10.1016/S0006-3495 (94)80644-7

[149] Zscherp C, Schlesinger R, Tittor J, Oesterhelt D, Heberle J. In situ determination of transient pK_a changes of internal amino acids of bacteriorhodopsin by using timeresolved attenuated total reflection Fourier-transform infrared spectroscopy. Proceedings of the National Academy of Sciences of the United States of America. 1999;**96**:5498-5503. DOI: 10.1073/pnas.96.10.5498

[150] Dioumaev AK, Brown LS, Needleman R, Lanyi JK. Coupling of the reisomerization of the retinal, proton uptake, and reprotonation of Asp-96 in the N photointermediate of bacteriorhodopsin. Biochemistry. 2001; **40**:11308-11317. DOI: 10.1021/bi011027d

[151] Luecke H, Schobert B, Cartailler JP, Richter HT, Rosengarth A, Needleman R, Lanyi JK. Coupling photoisomerization of retinal to directional transport in bacteriorhodopsin. Journal of Molecular Biology. 2000;**300**:1237-1255. DOI: 10.1006/jmbi.2000.3884

[152] Harris A, Ljumovic M, Bondar AN, Shibata Y, Ito S, Inoue K, Kandori H, Brown LS. A new group of eubacterial light-driven retinal-binding proton pumps with an unusual cytoplasmic proton donor. Biochimica et Biophysica Acta. 2015;**1847**:1518-1529. DOI: 10.1016/j.bbabio.2015.08.003

[153] Sudo Y, Yoshizawa S. Functional and photochemical characterization of a light-driven proton pump from the Gammaproteobacterium *Pantoea vagans*. Photochemistry and Photobiology. 2016; **92**:420-427. DOI: 10.1111/php.12585

[154] Dyukova T, Robertson B, Weetall H. Optical and electrical

characterization of bacteriorhodopsin films. Biosystems. 1997;**41**:91-98. DOI: 10.1016/S0303-2647(96)01665-6

[155] Wang T, Sessions AO, Lunde CS, Rouhani S, Glaeser RM, Duan Y, Facciotti MT. Deprotonation of D96 in bacteriorhodopsin opens the proton uptake pathway. Structure. 2013;**21**: 290-297. DOI: 10.1016/j.str.2012.12.018

[156] Haupts U, Tittor J, Bamberg E, Oesterhelt D. General concept for ion translocation by halobacterial retinal proteins: the isomerization/switch/ transfer (IST) model. Biochemistry. 1997;**36**:2-7. DOI: 10.1021/bi962014g

[157] Wang T, Facciotti MT, Duan Y. Schiff base switch II precedes the retinal thermal isomerization in the photocycle of bacteriorhodopsin. PLoS One. 2013;**8**: e69882. DOI: 10.1371/journal. pone.0069882

[158] Delaney JK, Schweiger U, Subramaniam S. Molecular mechanism of protein-retinal coupling in bacteriorhodopsin. Proceedings of the National Academy of Sciences of the United States of America. 1995;**92**: 11120-11124. DOI: 10.1073/ pnas.92.24.11120

[159] Iwamoto M, Sudo Y, Shimono K, Araiso T, Kamo N. Correlation of the O-Intermediate rate with the pK_a of Asp-75 in the dark, the counterion of the Schiff base of pharaonis phoborhodopsin (sensory rhodopsin II). Biophysical Journal. 2005;**88**:1215-1223. DOI: 10.1529/biophysj.104.045583

[160] Miyasaka T, Koyama K, Itoh I. Quantum conversion and image detection by a bacteriorhodopsin-based artificial photoreceptor. Science. 1992;**255**:342-344. DOI: 10.1126/science.255.5042.342

[161] Robertson B, Lukashev EP. Rapid pH change due to bacteriorhodopsin

measured with a tin-oxide electrode. Biophysical Journal. 1995;**68**:1507-1517. DOI: 10.1016/S0006-3495(95)80323-1

[162] Tamogami J, Kikukawa T, Miyauchi S, Muneyuki E, Kamo N. A tin oxide transparent electrode provides the means for rapid time-resolved pH measurements: application to photoinduced proton transfer of bacteriorhodopsin and proteorhodopsin. Photochemistry and Photobiology. 2009; **85**:578-589. DOI: 10.1111 / j.1751-1097.2008.00520.x

[163] Wu J, Ma D, Wang Y, Ming M, Balashov SP, Ding J. Efficient approach to determine the pK_a of the proton release complex in the photocycle of retinal proteins. The Journal of Physical Chemistry B. 2009;**113**:4482-4491. DOI: 10.1021/jp804838h

[164] Tamogami J, Sato K, Kurokawa S, Yamada T, Nara T, Demura M, Miyauchi S, Kikukawa T, Muneyuki E, Kamo N. Formation of M-like intermediates in proteorhodopsin in alkali solutions (pH ≥ ~8.5) where the proton release occurs first in contrast to the sequence at lower pH. Biochemistry. 2016;**55**:1036-1048. DOI: 10.1021/acs. biochem.5b01196

[165] Tsukamoto T, Inoue K, Kandori H, Sudo Y. Thermal and spectroscopic characterization of a proton pumping rhodopsin from an extreme thermophile. The Journal of Biological Chemistry. 2013;**288**:21581-21592. DOI: 10.1074/jbc. M113.479394

[166] Kanehara K, Yoshizawa S, Tsukamoto T, Sudo Y. A phylogenetically distinctive and extremely heat stable light-driven proton pump from the eubacterium *Rubrobacter xylanophilus* DSM 9941[T]. Scientific Reports. 2017;7:44427. DOI: 10.1038/srep44427

[167] Beppu K, Sasaki T, Tanaka KF, Yamanaka A, Fukazawa Y, Shigemoto R, Matsui K. Optogenetic countering of glial acidosis suppresses glial glutamate release and ischemic brain damage. Neuron. 2014;**81**:314-320. DOI: 10.1016/ j.neuron.2013.11.011

[168] Rost BR, Schneider F, Grauel MK, Wozny C, Bentz CG, Blessing A, Rosenmund T, Jentsch TJ, Schmitz D, Hegemann P, Rosenmund C. Optogenetic acidification of synaptic vesicles and lysosomes. Nature Neuroscience. 2015;**18**:1845-1852. DOI: 10.1038/nn.4161

[169] Hara KY, Wada T, Kino K, Asahi T, Sawamura N. Construction of photoenergetic mitochondria in cultured mammalian cells. Scientific Reports. 2013;**3**:1635. DOI: 10.1038/ srep01635

[170] Brown J, Behnam R, Coddington L, Tervo DGR, Martin K, Proskurin M, Kuleshova E, Park J, Phillips J, Bergs ACF, Gottschalk A, Dudman JT, Karpova AY. Expanding the optogenetics toolkit by topological inversion of rhodopsins. Cell. 2018;**175**: 1-10. DOI: 10.1016/j.cell.2018.09.026

[171] Inoue K, Ito S, Kato Y, Nomura Y, Shibata M, Uchihashi T, Tsunoda SP, Kandori H. A natural light-driven inward proton pump. Nature Communications. 2016;7:13415. DOI: 10.1038/ ncomms13415

[172] Inoue K, Tsunoda SP, Singh M, Tomida S, Hososhima S, Konno M, Nakamura R, Watanabe H, Bulzu PA, Banciu HL, Andrei AS, Uchihashi T, Ghai R, Béjà O, Kandori H. Schizorhodopsins: a family of rhodopsins from Asgard archaea that function as light-driven inward H⁺ pumps. Science Advances. 2020;**6**:eaaz2441. DOI: 10.1126/sciadv.aaz2441

[173] Altan N, Chen Y, Schindler M, Simon SM. Defective acidification in human breast tumor cells and implications for chemotherapy. Journal of Experimental Medicine. 1998;**187**: 1583-1598. DOI: 10.1084/jem.187.10.1583

[174] Gong Y, Duvvuri M, Krise JP. Separate roles for the Golgi apparatus and lysosomes in the sequestration of drugs in the multidrug-resistant human leukemic cell line HL-60. The Journal of Biological Chemistry. 2003;**278**: 50234-50239. DOI: 10.1074/jbc. M306606200

Chapter 8

Spatiotemporal Regulation of Cell–Cell Adhesions

Brent M. Bijonowski

Abstract

Cell–cell adhesions are fundamental in regulating multicellular behavior and lie at the center of many biological processes from embryoid development to cancer development. Therefore, controlling cell–cell adhesions is fundamental to gaining insight into these phenomena and gaining tools that would help in the bioartificial construction of tissues. For addressing biological questions as well as bottom-up tissue engineering the challenge is to have multiple cell types self-assemble in parallel and organize in a desired pattern from a mixture of different cell types. Ideally, different cell types should be triggered to self-assemble with different stimuli without interfering with the other and different types of cells should sort out in a multicellular mixture into separate clusters. In this chapter, we will summarize the developments in photoregulation cell–cell adhesions using non-neuronal optogenetics. Among the concepts, we will cover is the control of homophylic and heterophilic cell–cell adhesions, the independent control of two different types with blue or red light and the self-sorting of cells into distinct structures and the importance of cell–cell adhesion dynamics. These tools will give an overview of how the spatiotemporal regulation of cell–cell adhesion gives insight into their role and how tissues can be assembled from cells as the basic building block.

Keywords: optogenetics, cell–cell adhesion, differential adhesion hypothesis, reversible adhesion, subcellular resolution

1. Introduction

Cells adhere to the matrix and other cells around them, which fundamentally impacts their behavior. A thorough understanding of these adhesive interactions is also important to produce artificial tissues. Cell adhesions are formed by cell adhesion molecules on the cell surface such as integrins and cadherins which bind to the matrix and cadherins on neighboring cells, respectively [1]. These adhesion molecules transmit both physical and chemical signals between cells and their environment via the underlying cytoskeleton and intracellular signaling cascades [2].

1.1 Cell–cell adhesions

Cell–cell connections induce and receive biochemical signals and contractile forces from adjacent cells, and it is through theses stresses that cellular and tissue homeostasis is maintained [3]. The most abundant and well-studied cell–cell adhesion molecules are the cadherins. Cadherins such as E-cadherin, N-cadherin, and P-cadherin, consist of five extracellular domains with a calcium-binding site between each domain (**Figure 1**). The cell–cell adhesion is initiated by the cadherins on adjacent cells forming homophilic interactions via the exchange of β-strands between the first extracellular domains [4] and from here the cadherin signal is transmitted into the cell via an intracellular tail domain. Force-dependent conformation changes in cadherins lead to the recruitment of binding partners such as α-catenin, β-catenin, and vinculin thereby conveying the chemical signal to the intracellular actomyosin network. These ensuing biomechanical and biochemical cascades direct scaffolding proteins toward cellular pathways regulating division, survival, structural morphologies [5, 6] epithelial-mesenchymal transition (EMT), cell-sorting, and collective cell migration [7].

1.2 Spatiotemporal regulation of cell–cell adhesions

Altering the number of cellular adhesions is critical to many biological processes during tissue development and cancer progression. For instance, the interconnected nature of epithelial cells, which line the surface of organs, tissues, and blood vessels, designates their polarity, which is critical to their function. EMT takes place when epithelial cells lose the adhesions to other cells and therefore their basal-apical polarity. The resulting mesenchymal cell has increased cellular motility and invasiveness. This process takes place naturally to produce the mesoderm, one of the germ layers, during embryonic development [8, 9], pro-inflammatory wound healing [10], and during cancer cell metastasis [11–13].

Before the development of the germ layers, the embryonic stem cells in the inner mass of the blastocyst are largely epithelial in characteristic; however, during germ layer development, gastrulation, the epithelial-like cells undergo EMT to form the

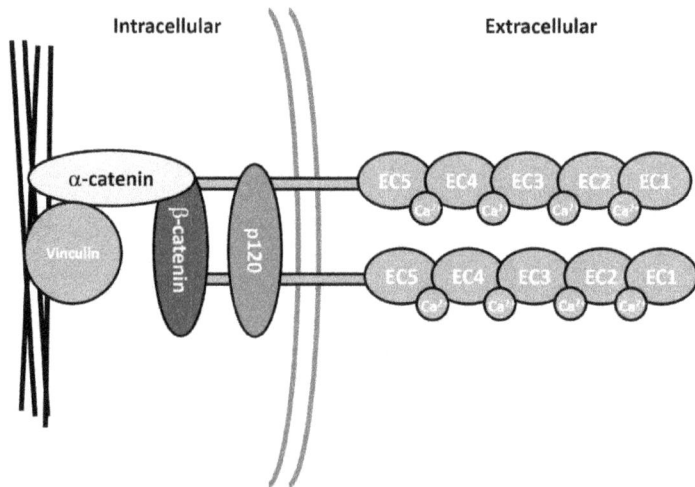

Figure 1.
E-cadherin dependent cell–cell adhesion. The E-cadherin consists of five extracellular domains, one transmembrane domain and an intracellular domain. During binding of two E-cadherin molecules the proteins p120, β-catenin, a-catenin, and vinculin get recruited to the intracellular domain leading to cytoskeletal adhesion and actomyosin based activation.

mesoderm. I*n vitro* culturing of embryonic stem cells or epiblast cell colonies, shows that they lose expression of E-cadherin, vimentin, and N-cadherin, thus giving rise to cells with a mesenchymal phenotype. The opposite of EMT, mesenchymal-epithelial transition (MET) also occurs naturally and can be seen in the procedure by which induced pluripotent stem cells are formed from fully differentiated cells. This process requires the transition from a cmesenchymal phenotype to an epithelial phenotype, and the activation of epithelial genes encoding epithelial cell junction proteins [8].

EMT extends to carcinomas as well, where a subpopulation of self-renewing cells, known as cancer stem cells, can efficiently generate new tumors. This can be seen in mammary carcinomas following the induction of EMT, which promotes the generation of clusters of invasive mammary gland cells [14]. The extent of these epithelial connections can also be seen in metastatic experiments involving the mammary cancer cell line MCF-7, which maintains an epithelial-like phenotype. In these experiments, MCF-7 is added on top of mammary endothelial cell sheets, and the invasiveness of MCF-7's was evaluated over increasing crossflow, it was revealed that the majority of MCF-7 cells could not form strong adhesions thereby failing to invade. Instead, the MCF-7 s remained rounded and rolled across the surface of the endothelial sheet [13].

Cadherin connections also guide cell migration through their intracellular connection to the cytoskeleton. For instance, in experiments examining the effect of cadherin adhesions in binary cell systems, it was revealed that single adhesions quickly recruit more cadherins to the initial contact site. Additionally, each recruited cadherin binds to the actin cytoskeleton preventing its depolymerization and enabling actomyosin-based mechanical signals [2, 15–17]. Additionally, cadherin-based stabilization of actin in migrating cells leads to *in situ* blebbing of the plasma membrane. These develop the leading edge for the cell, which in turn coordinates the migration of the cell [18]. In tissues with lots of interlocking cadherins, these effects lead to the development of leader cells, which migrate in front of the main body of follower cells. This is an event very common in angiogenesis, where sprouting endothelial cells lead to the development of new blood vessels [19].

1.3 Bottom-up tissue engineering

Another aspect for which controlled cell–cell adhesions are crucial is in bottom-up tissue engineering, in which single cells are organized into either planar or three-dimensional structures [20]. Since bottom-up engineering does not rely on external matrices to sequester the cells and instruct cellular arrangement the ability to spatio-temporally control the cell–cell connections is critical to building the desired structure. Techniques for creating bottom-up tissues include bioprinting, construction of cell sheets, and self-assembly of multicellular aggregates [20–23].

Self-assembled multicellular aggregates form by mixing multiple cell types such that microtissues with desired organization form. Generally, these structures form based on minimizing the potential internal energy resulting from cell–cell adhesions [24, 25]. Self-assembled aggregates have been used to construct multicell neuro-organoids comprised of cortical neural progenitor cells, endothelial cells, and mesenchymal stem cells. Different aggregates of each or a mix of two cells were first created in low-attachment 96-well plates. Following aggregate production, aggregates were then mixed to fuse the three cell populations. The resultant aggregate then sorted to form discrete layers within the aggregate. The cortical neural progenitor and endothelial cells developed into vascularized cortical brain tissue, while the mesenchymal stem cells took on a supportive role in the core of the aggregate [26]. With the ability to spatiotemporally control cell–cell adhesions it becomes possible to self-assemble cells together to produce more complex tissues that better recapitulate the *in vivo* structure.

1.4 Differential adhesion hypothesis

The cell sorting observed in tissues, self-assembled aggregates, and the developing embryoblast can be described by the differential adhesion hypothesis (DAH). The DAH explains cell sorting by comparing it to that of liquid mixtures, whereby the components (liquids or cells) arrange so that the internal free energy from cellular adhesions is reduced to a minimum to attain thermodynamic equilibrium [27–30]. Equilibrium is achieved via the active or passive motility of cells in the tissue rearranging with respect to each other to minimize stress and strain thereby limiting the internal energy [31]. Other aspects such as the cell's ability to round up to minimize their surface area, spreading of one cell over another, the fusion of two cellular aggregates, the sorting out behavior of mixed cell populations, and the hierarchy of the layering of two cell types further prove the analogy to liquid mixtures [31–33]. The DAH describes three different cases for multicellular assemblies in a mixture of two cell types (**Figure 2**) [30].

1.4.1 Intermixed

In this condition cells of type A and type B stay intermixed when the work of adhesion between the two cell populations (Wab) is higher than the work of cohesion of a single cell type (Wa and Wb) as this results in the maximal adhesion.

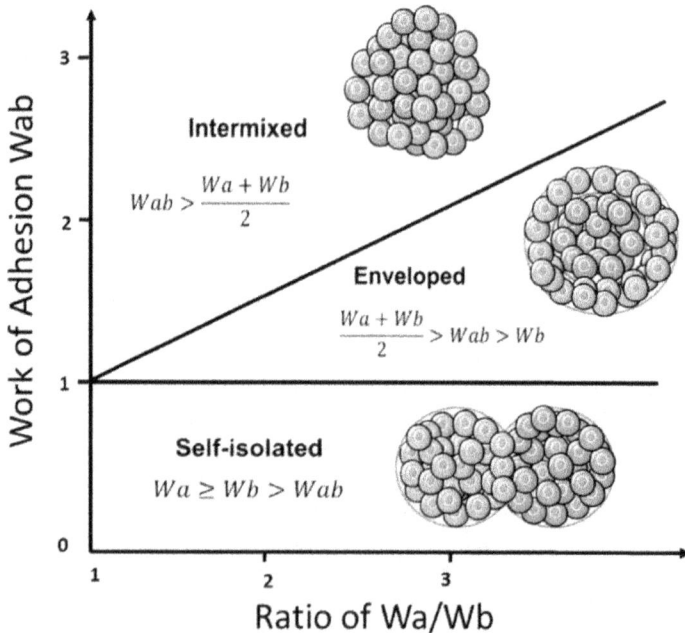

Figure 2.
Differential adhesion hypothesis (DAH). Different cell assemblies form at equilibrium depending on the work of adhesion between cells of type a (W_a), cells of type b (W_b) and cells of type a and type b (W_{ab}).

1.4.2 Enveloped

An enveloped arrangement of cells, occurs when one cell type is in the center and the secondary at its periphery. This arrangement forms when the average work of cohesion of cell type A and cell type B is greater than the work of adhesion between the two cell types and the work of cohesion of one cell type is smaller than the work of adhesion between the cell types. Herein, the cell type with the stronger cohesion, type A, forms the core and the less cohesive cell type, type B, surrounds this core.

1.4.3 Self-isolated

In a self-isolated system the two cell types form separate assemblies because the work of adhesion between the cell types is smaller than the work of cohesion within either population. In this case each cell type will self-isolate with no intermixing.

Numerous studies with cells expressing different types and amounts of cadherins have demonstrated these sorting schemes [34–36]. These studies show that the differences in homophylic and heterophilic cell–cell adhesions determine the outcome and the origin of these differences on adhesions are not important for the result.

2. Possible ways of controlling cell–cell adhesions

Currently, there are only a few tools for controlling cell–cell adhesion, which enable the studying of the underlying biology and for bottom-up tissue engineering. Important aspects to consider in the control of cell–cell adhesions are their specificity, their dynamics, and most importantly, their spatiotemporal regulation. The current approaches can be divided into two; the modification of the cell surface with chemically reactive groups and the genetic modification of cells to alter the expression of cell adhesion molecules [37].

In the following sections, we will discuss options of regulating cell–cell adhesions using reactive chemical groups and then consider photoregulation of cell–cell adhesions using light-responsive small molecules and finally optogenetic approaches. Light is especially advantageous as a trigger for cell–cell adhesions since light, as opposed to other stimuli like chemical inputs, temperature, redox etc., can be delivered with superior spatial and temporal control. Using a focused beam of light enables precise subcellular delivery, which can exclude the surrounding area. Secondly, light allows for temporal control as it can be turned on or off instantly making delivery or removal at the desired point instantaneous [38, 39].

2.1 Introduction of reactive groups to induce cell–cell adhesions

A general strategy for initiating user-controlled cell–cell interactions is to introduce reactive chemical groups on the cell surface. These chemical groups are not genetically coded and thus do not require genetic engineering to add them to the surface. Such chemical groups are introduced through the fusion of lipid vesicles containing the chemical reactive groups or through metabolic labeling with non-natural sugars bearing bioorthogonal functional groups with the cell [40]. For instance, complementarily reactive ketone and oxyamine groups or alkyne and azide groups can be introduced on the plasma membrane of cells [41]. Consequently, when cells with complementary reactive groups are mixed, the functional groups on the cell surfaces react and cells are connected through covalent bonds [42, 43]. In general, so-called click reactions, that take place in water, do not form toxic side products and do not interfere with other functional groups found in biomolecules. Alternatively, noncovalent interactions with high specificity can be used to form cell–cell adhesions. For this purpose, the binding of biotin to streptavidin [44–46] or the hybridization of complementary single-stranded DNA [47–49] is employed. DNA-based cell–cell adhesions open the possibility to form diverse structures with varying cell types and cellular connectivity owing to the high specificity of these interactions; however, DNA adhesions show limited reversibility making migratory sorting impossible, and covalent and strong noncovalent links between cells permanently glue them together [50].

2.2 Spatiotemporal control over cell–cell adhesions using light responsive small molecules

Light sensitive small molecules, such as nitrobenzenes and azobenzenes, have been used to control cell–cell adhesions in space and time. For example, light cleavable nitrobenzene groups can be introduced to oxyamine linkers at the cell surface. When this cell population is mixed with a second population of cells with a ketone group at the cell surface multicellular clusters formed. These cell cluster can then be broken up into single cells upon UV-light illumination since UV-light cleaves the nitrobenzyl moiety [51]. Such a photocleavable linker only allows for a single reversion of the cell–cell adhesions. To achieve cell–cell adhesions that can be switched on and off repeatedly a linker with a photoswitchable azobenzene group was developed. β-cyclodextrins can be clicked onto the surfaces of cells and when a divalent photoswitchable azobenzene (azo) linker (azo-PEG-azo) is added in the dark the cells will link together. This is because, in the dark, the trans configuration of the azobenzenes binds to the cyclodextrin moieties linking the cells together. Upon UV illumination the azobenzene switch to the cis conformation, which results in the release from the cyclodextrin and the dissociation of the cell–cell interactions. The azobenzene can then be switched back to the trans configuration with blue light illumination, thus allowing for the formation of new cell–cell adhesions [52]. These studies represent great advances in the field and allow for spatiotemporal control over cell–cell adhesions. However, the use of UV-light is damaging to DNA and therefore to cells, and secondly, the chemical modifications cannot be maintained over long periods of time. Thus, a system which utilizes biocompatible light and can be expressed over long times would be more beneficial to bottom-up tissue engineering since cell proliferation is a key component of any built tissue. For this purpose, a genetically engineered system, which allows for propagating the modification at the cell surface would be desirable.

2.3 Optogenetic control of cell–cell interactions

Cell–cell adhesions can be photoregulated by expressing bioartificial light-responsive proteins on the surfaces of cells as adhesion receptors. Numerous light-responsive proteins from algae, plants, bacteria, and engineered proteins change their conformation upon light illumination and bind to other proteins in a light-dependent manner through non-covalent protein–protein interactions [53–56]. In these optogenetic approaches, complementary light-dependent binding partners are expressed in the surfaces of different cell types by transfecting these proteins along with a plasma membrane localization sequence and a membrane anchoring sequence. Following translation, the localization sequence ensures that the protein is exported to the cell membrane where the extracellular portion operates as a bioartificial cell adhesion receptor [51, 52, 57]. For instance, the proteins Cryptochrome 2 (CRY2) from *Arabidopsis thaliana* and its blue light-dependent binding partner cryptochrome-interacting basic helix–loop–helix (CIBN) protein, were expressed on the surfaces of MDA-MB-231 cells, which do no form native cell–cell adhesion. When cells expressing CRY2 and CIBN at their surface are mixed and cultured in the dark, no cell–cell adhesions form similar to the parent MDA-MB-231 cell line. However, if these cells are cultured under blue light, the cells grow in clusters indicating the formation of cell–cell adhesions (**Figure 3**). Moreover, the cell–cell interactions formed under blue light can be reversed in the dark, allowing for repeated deconstruction and reconstruction with light-dependent control [58]. This optogenetic approach has the advantage that the cell–cell adhesions can be triggered with visible blue light, which is non-toxic to the cells and the cell surface modifications are passed on to daughter cells following cell splitting.

The large repertoire of photoswitchable protein–protein interactions allows for the formation of bioartificial cell–cell adhesions with different properties in terms of cell–cell adhesion mode, the light of color the adhesions responds to, reversion kinetics in the dark, and cell–cell adhesion dynamics [53–55].

In biology, cells can either interact with cells of their own type forming homophilic interactions or cells of another type forming heterophilic interactions.

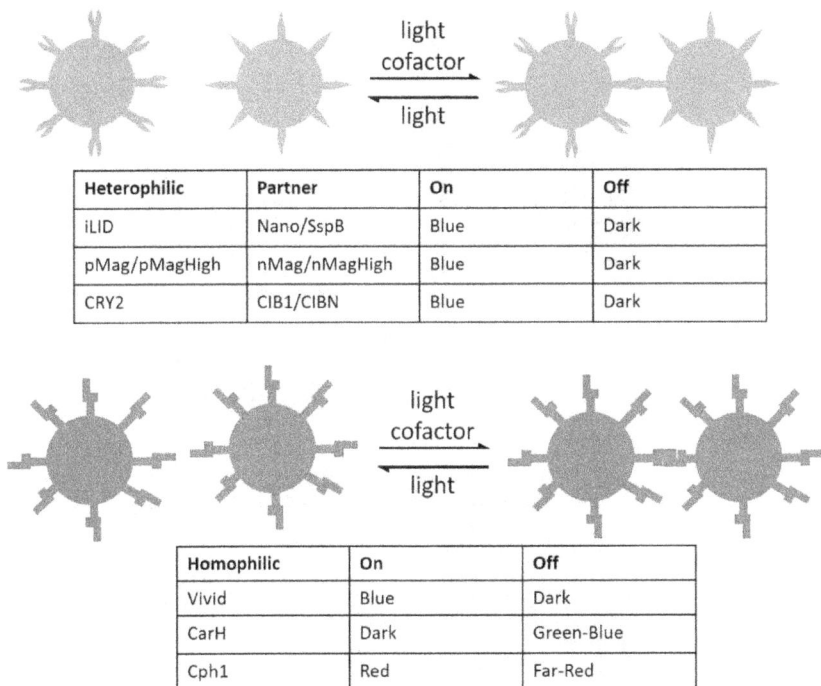

Heterophilic	Partner	On	Off
iLID	Nano/SspB	Blue	Dark
pMag/pMagHigh	nMag/nMagHigh	Blue	Dark
CRY2	CIB1/CIBN	Blue	Dark

Homophilic	On	Off
Vivid	Blue	Dark
CarH	Dark	Green-Blue
Cph1	Red	Far-Red

Figure 3.
Optogenetic proteins bind either in hetero or homophilic complexes. In heterophilic optogenetic systems an optogenetic protein undergoes conformational changes that enable the binding to a target protein. Homophilic optogenetic proteins also undergo conformation changes, but here a homomer is formed. iLID (improved light induced dimer), CRY2 (Cryptochrome 2), CIB1/N (cryptochrome-interacting basic helix–loop–helix/truncated), Cph1 (cyanobacterial phytochrome 1).

To obtain light-responsive homophilic cell–cell adhesion, proteins that homodi-merize under light are used as a mediator of cell–cell adhesion. For this purpose, the proteins Vivid, a member of the light oxygen voltage (LOV) domain from *Neurospora crassa*, and cyanobacterial phytochrome 1 (Cph1) from *Synechocysitis sp. PCC 6803* were used as these proteins homodimerize under blue and red light, respectively (**Figure 4**). Cells expressing Vivid at their plasma membrane form cell–cell adhesion exclusively when illuminated with blue light but not with red light. The reverse is true for cells expressing Cph1 at their cell surface, which only form cell–cell interactions under red light and not in the dark or under blue light. Similarly, the blue-green light-responsive protein, CarH from *Thermus thermophilus*, has been used to mediate homo-philic cell–cell interactions. The formation of a CarH homotetramer allows it to form cell–cell adhesions between cells expressing CarH on their surface in the dark [59].

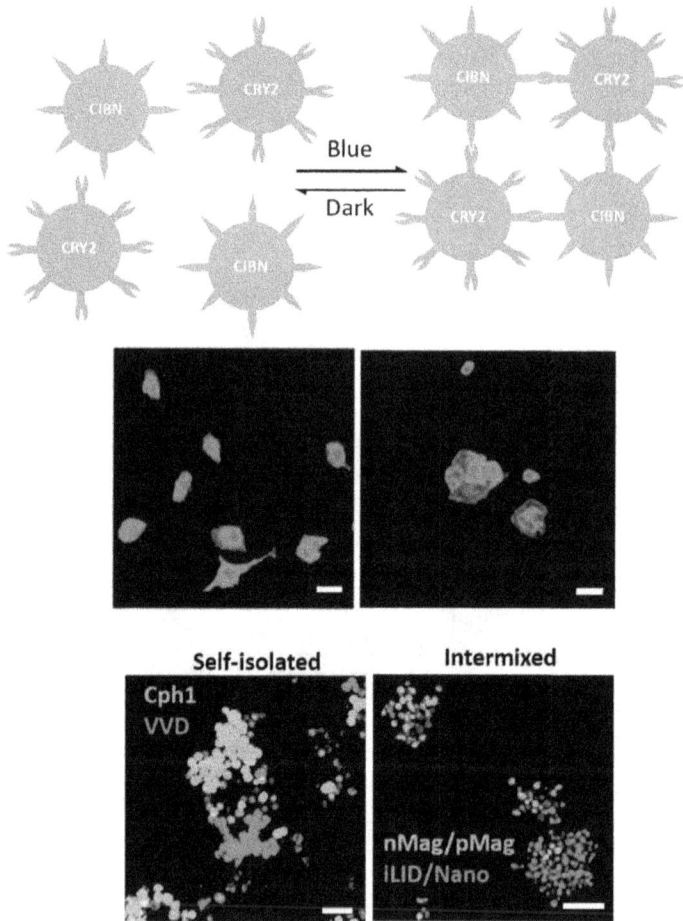

Figure 4.
Co-culture of optogenetic proteins results in cluster segregation. When colloidial particles are labled with the iLID/Nano, nMag/pMag, or nMagHigh/pMagHigh clusters of particles can be seen to form with respect to the kinetics of the system (adapted from Müller et al. [62]). In cellular systems utilizing the vivid (VVD) and Cph1 systems descrete clusters are observed rather than any intermixing (adapted from Rasoulinejad et al. [57]).

The CarH tetramer irreversibly degrades when exposed to blue-green light and hence the CarH based cell–cell adhesion can only be reversed once [59, 60].

Light responsive heterophilic cell–cell adhesions, can be achieved by proteins that heterodimerize under light to form cell–cell adhesions. For this purpose, different heterodimerization pairs that form under blue light and reverse in the dark were used. These include the binding of the improved light-induced dimer (iLID) to Nano [61], the binding of the Vivid-based proteins nMag and pMag and the previously-described binding of CRY2 to CIBN. These different protein pairs provide different interaction strengths, reversion kinetics in the dark, and protein–protein dynamics.

2.4 Cell–cell adhesion dynamics dictate the structure of multicellular assemblies

The assembly of multicellular structures does not just depend on the strength of the underlying cell–cell adhesions but also their dynamics. If cell–cell adhesions are dynamic, meaning that formed protein–protein interactions constantly form and disassemble within the chemical equilibrium, cells can move with respect to each other and maximize the number of adhesive contacts they form. This scenario is observed in mixtures of iLID and Nano expressing cells, which assembled into spherical and compact clusters. If cell–cell adhesions are not dynamic, meaning that once protein–protein interactions form that they do not reverse, cells stick to the first cell they meet and cannot move to areas with potential higher numbers of adhesions. For example, mixtures of nMagHigh and pMagHigh or nMag and pMag expressing cells assemble into ramified branched structures, which are kinetically trapped. Optogenetics allows for the altering of the dynamics of the cell–cell adhesion by turning light on and off. The ramified structures formed with nMag and pMag cells could then be converted into compact spheres under pulsed illumination (5 min on, 5 min off), allowing the adhesions to dissipate and the cells to move.

2.5 Regulation of cell sorting using photoswitchable cell–cell adhesions

Different types of photoswitchable cell–cell adhesions can be mixed to obtain cell sorting within multicellular mixtures and organize cells as predicted by the DAH. For example, when cells expressing Vivid or Cph1 at their cell surface were mixed and illuminated with either blue or red-light clusters of cells with Vivid or Cph1 cells formed, respectively. When both blue and red light was used self-isolated clusters containing either Vivid or Cph1 cells were observed (**Figure 4**) [57]. That means that the adhesive force for Vivid and Cph1 is lower than that for the homodimers formed for each system due to the specific protein–protein interactions. Similarly, also different pairs of heterophilic cell–cell adhesions can be used to achieve self-sorting in mixtures containing four different cell types. In mixtures of iLID, Nano, nMag, and pMag expressing cells, two types of multicellular aggregates assembled each containing one of the protein pairs (iLID/Nano or nMag/pMag) [62]. It should be noted that cell sorting is only possible if the system is under thermodynamic control and is not observed if kinetically trapped structures form. Therefore, mixtures of iLID, Nano, nMagHigh and pMagHigh do not sort into distinct clusters.

2.6 Photoswitchable cell–cell adhesions controlling cellular function

Cell–cell adhesions play an important role in many cellular functions, and the adhesions resulting from the optogenetic proteins are no different. Using CarH based homophilic cell–cell adhesions, the spatiotemporal control of migrations was assessed by measuring the rate and the morphology of cells migrating during a wound-healing assay. The spatiotemporal element was carried out by illuminating discrete sections to depolymerize the cell–cell adhesion. Cells with intact CarH adhesions in the dark showed significantly enhanced migratory potential compared to cells illuminated with green light, which dissociate the cell–cell adhesions. This was characterized by cells remaining together and thus migrating as a single cell wall resulting in faster migration. Cells that were illuminated with blue-green light broke away from the migratory front and engaged in random walking resulting in a slower overall migration rate [59].

Additionally, spatiotemporal control of the cell–cell adhesion complex has been shown in experiments where the β-catenin binding domains on E-cadherin and α-catenin have been replaced with the Halo and SNAP tags, respectively. The Halo/SNAP system incorporates the UV-light photocleavable small molecule Ha-pl-BG, so adhesions can be reversed upon UV illumination. This system was then applied to MDA-MB-468, which do not express endogenous E-cadherin to assess the efficacy of the system. Using the system cell–cell adhesions could only be observed when the cofactor was present and were degraded rapidly under UV-light. To illustrate the spatiotemporal control, A431 cells, with knocked out α-catenin, were labeled with the Halo/SNAP system and cultured overnight to initiate connections between cells. Specific adhesions between cells were then targeted and illuminated with UV-light. Only the targeted connections were degraded leaving the other connections intact.

3. Conclusion

The spatiotemporal nature of cadherin-based cell–cell adhesions enables cells to self-sort, assemble into tissues, or can lead to cellular differentiation. However, these adhesions cannot be exogenously controlled, and as such make the construction of bottom-up tissues difficult to manage. Chemical means for binding cell membranes together are too rigid and offer limited reversibility. There is also a lack of spatio-temporal control. However, light is non-invasive, highly biocompatible, and can be delivered in a spatiotemporal fashion. Through the delivery of optogenetic proteins to the cell membrane, the construction of spatiotemporal cell–cell adhesions has been achieved. These proteins can respond to a wide range of wavelengths enabling the use of multiple pairs to construct larger structures, form reversible adhesions, and offer superior kinetics to other adhesion methods.

Acknowledgements

This work was funded by the European Research Council ERC Starting Grant ARTIST (# 757593).

Conflict of interest

The authors have no conflicts to disclose.

Author details

Brent M. Bijonowski
Institute for Physiological Chemistry and Pathobiochemistry, University of Münster, Münster, Germany

*Address all correspondence to: bijonows@uni-muenster.de

IntechOpen

References

[1] APLIN A, HOWE A, JULIANO R. Cell adhesion molecules, signal transduction and cell growth. Curr Opin Cell Biol. 1999;11(6):737-44. Available from: http://dx.doi.org/10.1016/s0955-0674(99)00045-9

[2] Han MKL, de Rooij J. Resolving the cadherin–F-actin connection. Nat Cell Biol. 2016;19(1):14-6. Available from: http://dx.doi.org/10.1038/ncb3457

[3] Nollet F, Kools P, van Roy F. Phylogenetic analysis of the cadherin superfamily allows identification of six major subfamilies besides several solitary members 1 1Edited by M. Yaniv. J Mol Biol. 2000;299(3):551-72. Available from: http://dx.doi.org/10.1006/jmbi.2000.3777

[4] Shapiro L, Weis WI. Structure and biochemistry of cadherins and catenins. Cold Spring Harb Perspect Biol. 2009 Sep;1(3):a003053–a003053. Available from: https://pubmed.ncbi.nlm.nih.gov/20066110

[5] Abdallah BM, Kassem M. Human mesenchymal stem cells: from basic biology to clinical applications. Gene Ther. 2007;15(2):109-16. Available from: http://dx.doi.org/10.1038/sj.gt.3303067

[6] Olmer R, Lange A, Selzer S, Kasper C, Haverich A, Martin U, et al. Suspension culture of human pluripotent stem cells in controlled, stirred bioreactors. Tissue Eng Part C Methods. 2012/06/04. 2012 Oct;18(10):772-84. Available from: https://pubmed.ncbi.nlm.nih.gov/22519745

[7] Klezovitch O, Vasioukhin V. Cadherin signaling: keeping cells in touch. F1000Research. 2015 Aug 12;4(F1000 Faculty Rev):550. Available from: https://pubmed.ncbi.nlm.nih.gov/26339481

[8] Lamouille S, Xu J, Derynck R. Molecular mechanisms of epithelial-mesenchymal transition. Vol. 15, Nature Reviews Molecular Cell Biology. Nature Publishing Group; 2014. p. 178-96. Available from: www.nature.com/reviews/molcellbio

[9] Kong D, Li Y, Wang Z, Sarkar F. Cancer Stem Cells and Epithelial-to-Mesenchymal Transition (EMT)-Phenotypic Cells: Are They Cousins or Twins? Cancers (Basel). 2011 Feb 21;3(1):716-29. Available from: http://www.mdpi.com/2072-6694/3/1/716

[10] Borthwick LA, McIlroy EI, Gorowiec MR, Brodlie M, Johnson GE, Ward C, et al. Inflammation and Epithelial to Mesenchymal Transition in Lung Transplant Recipients: Role in Dysregulated Epithelial Wound Repair. Am J Transplant. 2010 Mar 1;10(3):498-509. Available from: http://doi.wiley.com/10.1111/j.1600-6143.2009.02953.x

[11] Nakaya Y, Sheng G. EMT in developmental morphogenesis. Vol. 341, Cancer Letters. Elsevier Ireland Ltd; 2013. p. 9-15.

[12] Chen T, You Y, Jiang H, Wang ZZ. Epithelial-mesenchymal transition (EMT): A biological process in the development, stem cell differentiation, and tumorigenesis. J Cell Physiol. 2017 Dec 1;232(12):3261-72. Available from: http://doi.wiley.com/10.1002/jcp.25797

[13] Thompson TJ, Han B. Analysis of adhesion kinetics of cancer cells on inflamed endothelium using a microfluidic platform. Biomicrofluidics. 2018 Jul 1;12(4):042215. Available from: http://aip.scitation.org/doi/10.1063/1.5025891

[14] Mani SA, Guo W, Liao MJ, Eaton EN, Ayyanan A, Zhou AY, et al. The Epithelial-Mesenchymal Transition Generates Cells with Properties of Stem Cells. Cell. 2008 May 16;133(4):704-15.

[15] Han MKL, de Rooij J. Converging and Unique Mechanisms of Mechanotransduction at Adhesion Sites. Trends Cell Biol. 2016;26(8):612-23. Available from: http://dx.doi.org/10.1016/j.tcb.2016.03.005

[16] Bertocchi C, Wang Y, Ravasio A, Hara Y, Wu Y, Sailov T, et al. Nanoscale architecture of cadherin-based cell adhesions. Nat Cell Biol. 2016/12/19. 2017 Jan;19(1):28-37. Available from: https://pubmed.ncbi.nlm.nih.gov/27992406

[17] Engl W, Arasi B, Yap LL, Thiery JP, Viasnoff V. Actin dynamics modulate mechanosensitive immobilization of E-cadherin at adherens junctions. Nat Cell Biol. 2014;16(6):584-91. Available from: http://dx.doi.org/10.1038/ncb2973

[18] Grimaldi C, Schumacher I, Boquet-Pujadas A, Tarbashevich K, Vos BE, Bandemer J, et al. E-cadherin focuses protrusion formation at the front of migrating cells by impeding actin flow. Nat Commun. 2020 Dec 1;11(1):1-15. Available from: https://doi.org/10.1038/s41467-020-19114-z

[19] Yamada KM, Sixt M. Mechanisms of 3D cell migration. Vol. 20, Nature Reviews Molecular Cell Biology. Nature Research; 2019. p. 738-52. Available from: https://pubmed.ncbi.nlm.nih.gov/31582855/

[20] Elbert DL. Bottom-up tissue engineering. Curr Opin Biotechnol. 2011/04/27. 2011 Oct;22(5):674-80. Available from: https://pubmed.ncbi.nlm.nih.gov/21524904

[21] Jakab K, Norotte C, Marga F, Murphy K, Vunjak-Novakovic G, Forgacs G. Tissue engineering by self-assembly and bio-printing of living cells. Biofabrication. 2010/06/02. 2010 Jun;2(2):22001. Available from: https://pubmed.ncbi.nlm.nih.gov/20811127

[22] Haraguchi Y, Shimizu T, Sasagawa T, Sekine H, Sakaguchi K, Kikuchi T, et al. Fabrication of functional three-dimensional tissues by stacking cell sheets in vitro. Nat Protoc. 2012;7(5): 850-8. Available from: http://dx.doi.org/10.1038/nprot.2012.027

[23] Saleh FA, Genever PG. Effects of endothelial cells on human mesenchymal stem cell activity in a three-dimensional in vitro model Pharmacological effects of medicinal plants View project. Eur Cells Mater. 2011;22:242-57. Available from: www.ecmjournal.org

[24] Athanasiou KA, Eswaramoorthy R, Hadidi P, Hu JC. Self-organization and the self-assembling process in tissue engineering. Annu Rev Biomed Eng. 2013/05/20. 2013;15:115-36. Available from: https://pubmed.ncbi.nlm.nih.gov/23701238

[25] Bijonowski BM, Yuan X, Jeske R, Li Y, Grant SC. Cyclical aggregation extends in vitro expansion potential of human mesenchymal stem cells. Sci Rep. 2020 Dec 1;10(1):1-10. Available from: https://doi.org/10.1038/s41598-020-77288-4

[26] Song L, Yuan X, Jones Z, Griffin K, Zhou Y, Ma T, et al. Assembly of Human Stem Cell-Derived Cortical Spheroids and Vascular Spheroids to Model 3-D Brain-like Tissues. Sci Rep. 2019 Dec 1;9(1):1-16. Available from: https://doi.org/10.1038/s41598-019-42439-9

[27] Steinberg MS. Mechanism of Tissue Reconstruction by Dissociated Cells, II : Time-Course of Events inine Scene-Utilization of Nitrogen Compounds by Unicellular Algae. Science. 1962;137 (X 127):762-3.

[28] Steinberg MS. ON THE MECHANISM OF TISSUE RECONSTRUCTION BY DISSOCIATED CELLS, III. FREE ENERGY RELATIONS AND THE REORGANIZATION OF FUSED, HETERONOMIC TISSUE FRAGMENTS. Proc Natl Acad Sci U S A. 1962 Oct;48(10):1769-76. Available from: https://pubmed.ncbi.nlm.nih.gov/16591009

[29] Steinberg MS. On the mechanism of tissue reconstruction by dissociated cells. I. Population kinetics, differential adhesiveness. and the absence of directed migration. Proc Natl Acad Sci U S A. 1962 Sep 15;48(9):1577-82. Available from: https://pubmed.ncbi.nlm.nih.gov/13916689

[30] Steinberg MS. Reconstruction of Tissues by Dissociated Cells. Science (80-). 1963;141(3579):401-8. Available from: http://dx.doi.org/10.1126/science.141.3579.401

[31] Foty RA, Steinberg MS. Differential adhesion in model systems. Wiley Interdiscip Rev Dev Biol. 2013;2(5):631-45. Available from: http://dx.doi.org/10.1002/wdev.104

[32] Palsson E. A 3-D model used to explore how cell adhesion and stiffness affect cell sorting and movement in multicellular systems. J Theor Biol. 2008;254(1):1-13. Available from: http://dx.doi.org/10.1016/j.jtbi.2008.05.004

[33] Tsai A-C, Liu Y, Yuan X, Ma T. Compaction, fusion, and functional activation of three-dimensional human mesenchymal stem cell aggregate. Tissue Eng Part A. 2015/03/20. 2015 May;21(9-10):1705-19. Available from: https://pubmed.ncbi.nlm.nih.gov/25661745

[34] Duguay D, Foty RA, Steinberg MS. Cadherin-mediated cell adhesion and tissue segregation: qualitative and quantitative determinants. Dev Biol. 2003;253(2):309-23. Available from: http://dx.doi.org/10.1016/s0012-1606(02)00016-7

[35] Nose A, Nagafuchi A, Takeichi M. Expressed recombinant cadherins mediate cell sorting in model systems. Cell. 1988;54(7):993-1001. Available from: http://dx.doi.org/10.1016/0092-8674(88)90114-6

[36] Katsamba P, Carroll K, Ahlsen G, Bahna F, Vendome J, Posy S, et al. Linking molecular affinity and cellular specificity in cadherin-mediated adhesion. Proc Natl Acad Sci U S A. 2009/06/24. 2009 Jul 14;106(28):11594-9. Available from: https://pubmed.ncbi.nlm.nih.gov/19553217

[37] Stephan MT, Irvine DJ. Enhancing Cell therapies from the Outside In: Cell Surface Engineering Using Synthetic Nanomaterials. Nano Today. 2011 Jun 1;6(3):309-25. Available from: https://pubmed.ncbi.nlm.nih.gov/21826117

[38] Xu D, Ricken J, Wegner S V. Turning Cell Adhesions ON or OFF with High Spatiotemporal Precision Using the Green Light Responsive Protein CarH.; Available from: https://doi.org/10.1002/chem.202001238

[39] Karunarathne WKA, O'Neill PR, Gautam N. Subcellular optogenetics - controlling signaling and single-cell behavior. J Cell Sci. 2014/11/28. 2015 Jan 1;128(1):15-25. Available from: https://pubmed.ncbi.nlm.nih.gov/25433038

[40] Johnson HE, Goyal Y, Pannucci NL, Schüpbach T, Shvartsman SY, Toettcher JE. The Spatiotemporal Limits of Developmental Erk Signaling. Dev Cell. 2017 Jan 23;40(2):185-92. Available from: https://pubmed.ncbi.nlm.nih.gov/28118601

[41] Dutta D, Pulsipher A, Luo W, Yousaf MN. Synthetic Chemoselective Rewiring of Cell Surfaces: Generation of Three-Dimensional Tissue Structures. J Am Chem Soc. 2011;133(22):8704-13. Available from: http://dx.doi.org/10.1021/ja2022569

[42] O'Brien PJ, Luo W, Rogozhnikov D, Chen J, Yousaf MN. Spheroid and Tissue Assembly via Click Chemistry in Microfluidic Flow. Bioconjug Chem. 2015;26(9):1939-49. Available from: http://dx.doi.org/10.1021/acs.bioconjchem.5b00376

[43] Koo H, Choi M, Kim E, Hahn SK, Weissleder R, Yun SH. Bioorthogonal Click Chemistry-Based Synthetic Cell Glue. Small. 2015/11/19. 2015 Dec 22;11(48):6458-66. Available from: https://pubmed.ncbi.nlm.nih.gov/26768353

[44] Sarkar D, Spencer JA, Phillips JA, Zhao W, Schafer S, Spelke DP, et al. Engineered cell homing. Blood. 2011/10/27. 2011 Dec 15;118(25):e184-91. Available from: https://pubmed.ncbi.nlm.nih.gov/22034631

[45] Wang B, Song J, Yuan H, Nie C, Lv F, Liu L, et al. Multicellular Assembly and Light-Regulation of Cell-Cell Communication by Conjugated Polymer Materials. Adv Mater. 2013;26(15):2371-5. Available from: http://dx.doi.org/10.1002/adma.201304593

[46] De Bank PA, Hou Q, Warner RM, Wood I V, Ali BE, MacNeil S, et al. Accelerated Formation of multicellular §-D Structures by Cell-to-Cell-Linking. Biortechnology Bioeng. 2007;97(6):1460-9.

[47] Gartner ZJ, Bertozzi CR. Programmed assembly of 3-dimensional microtissues with defined cellular connectivity. Proc Natl Acad Sci U S A.

2009/03/09. 2009 Mar 24;106(12):4606-10. Available from: https://pubmed.ncbi.nlm.nih.gov/19273855

[48] Taylor MJ, Husain K, Gartner ZJ, Mayor S, Vale RD. A DNA-Based T Cell Receptor Reveals a Role for Receptor Clustering in Ligand Discrimination. Cell. 2017 Mar 23;169(1):108-119.e20.

[49] Togo S, Sato K, Kawamura R, Kobayashi N, Noiri M, Nakabayashi S, et al. Quantitative evaluation of the impact of artificial cell adhesion via DNA hybridization on E-cadherin-mediated cell adhesion. APL Bioeng. 2020 Mar 1;4(1):016103. Available from: http://aip.scitation.org/doi/10.1063/1.5123749

[50] Xiong X, Liu H, Zhao Z, Altman MB, Lopez-Colon D, Yang CJ, et al. DNA aptamer-mediated cell targeting. Angew Chem Int Ed Engl. 2012/12/11. 2013 Jan 28;52(5):1472-6. Available from: https://pubmed.ncbi.nlm.nih.gov/23233389

[51] Luo W, Pulsipher A, Dutta D, Lamb BM, Yousaf MN. Remote control of tissue interactions via engineered photo-switchable cell surfaces. Sci Rep. 2014 Sep 10;4:6313. Available from: https://pubmed.ncbi.nlm.nih.gov/25204325

[52] Shi P, Ju E, Yan Z, Gao N, Wang J, Hou J, et al. Spatiotemporal control of cell–cell reversible interactions using molecular engineering. Nat Commun. 2016 Oct 6;7:13088. Available from: https://pubmed.ncbi.nlm.nih.gov/27708265

[53] Fenno L, Yizhar O, Deisseroth K. The development and application of optogenetics. Annu Rev Neurosci. 2011;34:389-412. Available from: https://pubmed.ncbi.nlm.nih.gov/21692661

[54] Deisseroth K. Optogenetics - Method of the Year. Nat Methods. 2010;8(1):1-4.

[55] Chatelle C, Ochoa-Fernandez R, Engesser R, Schneider N, Beyer HM, Jones AR, et al. A Green-Light-Responsive System for the Control of Transgene Expression in Mammalian and Plant Cells. ACS Synth Biol. 2018 May 18;7(5):1349-58. Available from: https://pubmed.ncbi.nlm.nih.gov/29634242/

[56] Glantz ST, Carpenter EJ, Melkonian M, Gardner KH, Boyden ES, Wong GK-S, et al. Functional and topological diversity of LOV domain photoreceptors. Proc Natl Acad Sci. 2016 Mar;113(11):E1442 LP-E1451.

[57] Rasoulinejad S, Müller M, Nzigou Mombo B, Wegner S V. Orthogonal Blue and Red Light Controlled Cell-Cell Adhesions Enable Sorting-out in Multicellular Structures. ACS Synth Biol. 2020 Aug 21;9(8):2076-86. Available from: /pmc/articles/PMC7757848/?report=abstract

[58] Yüz SG, Rasoulinejad S, Mueller M, Wegner AE, Wegner S V. Blue Light Switchable Cell–Cell Interactions Provide Reversible and Spatiotemporal Control Towards Bottom-Up Tissue Engineering. Adv Biosyst. 2019;3(4):1800310. Available from: http://dx.doi.org/10.1002/adbi.201800310

[59] Nzigou Mombo B, Bijonowski BM, Rasoulinejad S, Mueller M, Wegner S V. Spatiotemporal Control Over Multicellular Migration Using Green Light Reversible Cell–Cell Interactions. Adv Biol. 2021 Jan 14;2000199. Available from: https://onlinelibrary.wiley.com/doi/10.1002/adbi.202000199

[60] Wang X, Chen X, Yang Y. Spatiotemporal control of gene expression by a light-switchable transgene system. Nat Methods. 2012;9(3):266-9. Available from: http://dx.doi.org/10.1038/nmeth.1892

[61] Guntas G, Hallett RA, Zimmerman SP, Williams T, Yumerefendi H, Bear JE, et al. Engineering an improved light-induced dimer (iLID) for controlling the localization and activity of signaling proteins. Proc Natl Acad Sci U S A. 2014/12/22. 2015 Jan 6;112(1):112-7. Available from: https://pubmed.ncbi.nlm.nih.gov/25535392

[62] Müller M, Rasoulinejad S, Garg S, Wegner S V. The Importance of Cell-Cell Interaction Dynamics in Bottom-Up Tissue Engineering: Concepts of Colloidal Self-Assembly in the Fabrication of Multicellular Architectures. Nano Lett. 2020 Apr 8;20(4):2257-63. Available from: https://pubs.acs.org/sharingguidelines

www.ingramcontent.com/pod-product-compliance
Lightning Source LLC
Chambersburg PA
CBHW081559190326
41458CB00015B/5655